GEOGR...

John Dawson is currently Reader in Geography at the
University of Wales. He has also held visiting positions
at the University of Western Australia, the Australian
National University, and the Florida State University.
His particular interests are in the geography of market-
ing, and in the application of computers to geographical
research and to teaching. Much of his recent work has
been related to the development and planning of shopping centres and hypermarkets. Dr Dawson has published several books on these topics, and many papers
concerned with the teaching of geography.

TEACH YOURSELF BOOKS

GEOGRAPHY

John A. Dawson

TEACH YOURSELF BOOKS

Hodder and Stoughton

First printed 1983

Copyright © 1983
John A. Dawson

ISBN 0 340 26827 1

Printed in Great Britain for
Hodder and Stoughton Educational,
a division of Hodder and Stoughton Ltd,
Mill Road, Dunton Green, Sevenoaks, Kent,
by Richard Clay (The Chaucer Press) Ltd,
Bungay, Suffolk.
Photoset by Rowland Phototypesetting Ltd,
Bury St Edmunds, Suffolk.

Contents

methods. Sample size. Economic surveys. Survey analysis.

PART THREE *Patterns in the Landscape*

PART FOUR *Explaining the Landscape*

Acknowledgments

Many people have helped in the preparation of this book and although any errors are mine I would like to thank particularly: Trevor Harris for his cartographic skills; Noreen and David Kay for their help in clearing some of my hazy concepts; an anonymous referee whose constructive criticism stimulated restructuring of certain sections; Margaret Jones and Maureen Hunwicks for their cheerfulness in the tedious typing task; and finally Jo Dawson who has improved the text in immeasurable ways. Without them this book would not have been possible.

John Dawson

Introduction:
Approaches to Geography

'Geography is the science which describes the surface of the earth:
the situation, extent, boundaries, and divisions of its different
countries, and the manners, customs, and political relations of its
inhabitants . . .'

> G. Roberts, *The Elements of Modern Geography* (1829)

'Geographers are concerned with the structure and interaction of
two major systems: the ecological system that links man and his
environment, and a spatial system that links one region with
another in a complex interchange of flows . . .'

> P. Haggett, *Geography: A Modern Synthesis* (1972)

Such quotations show how the content of geography has changed in
the last 150 years, but today geographers continue to debate what
geography *is*. Any attempt in a short book such as this to cover the
full breadth of geographical studies, or to give consideration to all
the approaches and philosophies of geography, is necessarily
doomed to failure. This book therefore reflects one, widely
accepted, view of contemporary geography.

For many decades geographers described countries and regions,
relating together the various facets of the environment to produce a
regional description, and so became experts on particular continents or countries. In recent decades a redefinition of geography has
taken place and many geographers are now more concerned with
the analysis of landscape types and features rather than confining
study to a region or country. Geographers have become specialists,
for example, in the study of urban landscapes; or arid landscapes; or

glacial landscapes; or economic landscapes. Even greater degrees of specialism may be seen in books and journals written by geographers. In urban geography, for example, they may focus only on industrial aspects of the urban landscape; or on urban crime; or on shopping centres in the city; or on slum housing. The range of specialisms is now very great, but the unifying theme is the attempt to interpret the landscape – what geographers often call *spatial structure*. Interpretation involves not only description of what landscapes look like, but also analysis of the patterns and associations in the landscape, explanation of the way landscape features change and the processes which govern these changes, and finally study of the ways man may control and use the processes creating the landscape. This alternative approach to geography, which has come about in recent decades, draws on the expertise of earlier approaches, uses this alongside borrowings from other subjects such as geology and economics, and develops new ideas and approaches to understanding the landscapes we see around us.

The organisation of this book is based on a belief that geographers have first to describe landscape, then analyse it, searching for patterns and associations, before explaining how it has come about. Chapters 1, 2 and 3, comprising Part I, consider approaches to describing landscape and draw on both long established and newer concepts. As geography has been redefined there has sometimes been a tendency to discard all the older concepts, but some longer established approaches are still valuable. Unlike mathematics or economics, geography does not have a set of logical theorems which provide theoretical foundations to the subject, and so it is a cumulative discipline constantly reassessing earlier concepts and applying them to new ideas. The awareness of the interactions amongst landscape features, and the need to describe interrelationships, for example, has led to the use of concepts of systems as considered in Chapter 3.

In order to describe landscape the geographer has tools of the trade. Three such tools are considered in Part II: maps, air photographs and surveys. These do not complete the geographer's tool box, as statistical and laboratory techniques are also frequently used, but the aim in this book is to review the basic tools. There are other books concerned specifically with geographical techniques, and some guidance on these is given at the end of the book. The use of these techniques provides the geographer with the data essential for analysis and also allows the geographer to search for order and patterns in these data.

Chapters 7 to 11 in Part III are concerned with patterns in the landscape both on a global scale, in the consideration of world patterns of climate and economic development, and on a regional scale, in, for example, studies of land-use patterns in cities. The patterns may be of regions, or flows, or points within the landscape. Although regions, flows and points may be closely related and interconnecting, the geographer often finds it useful, as a step in landscape interpretation, to separate them from each other.

Part IV takes landscape interpretation to the stages of explanation and possible control, and considers some of the types of process responsible for landscape change. Cyclical change in Chapter 12, change in a progression of stages in Chapter 13, diffusion in Chapter 14, and events in Chapter 15, cover the major types of landscape change. Again the aim is not to be comprehensive, nor to cover the full breadth of geography.

The final two chapters consider man as an active agent in landscape and environmental change. The effect of man on the landscape is implied in many of the earlier chapters. Man has an all-pervasive impact on contemporary landscapes, and some would argue that even by studying landscape we affect it.

The framework of the book is therefore an initial part considering definitions and ways of describing landscape; a second part considering the geographer's tools; a third part dealing with patterns in the landscape as revealed by the approaches of Part I and the techniques described in Part II; and a final part which reviews ways of explaining and fully understanding the patterns in the landscape. With such a framework the book is intended to be free-standing, but supplemented by a range of more specialist material providing case studies and greater depth of analysis for subjects of specific interest to the reader. The aim is to introduce the reader to some of the ideas in contemporary geography, to erase the 'capes and bays' image of geography still prevalent amongst non-geographers, and most importantly, to help those who are not content just to look at the landscape around them but who would like to understand what they see.

This is not a textbook for the public examination system of any country. Many such texts exist and there is nothing to be gained by providing another. Reading lists at the end of the book provide reference to textbooks. But do remember you cannot 'Teach Yourself Geography' simply by reading this book or the others listed. Understanding geography also involves looking at landscapes and interpreting what you see. Only in this way can the ideas and concepts in books be applied to the environment in which we live.

PART ONE

Describing the Landscape

1

Contrasts in the Landscape

As travellers we look at landscape and become aware of differences between places. There are variations within rural landscapes, contrasts between town and countryside, differences from one town to the next, and distinctions between suburbs and town centre. At the end of a journey the environment we see, smell, hear and use, is frequently quite different from our surroundings when we set out. We are conscious of the difference, but are not always able to pinpoint it immediately. Sometimes we choose to visit places that are attractive to us just because they are different from home. So Britons visit the bulb fields of Lincolnshire, the hills of the Lake District, the coast of the South West peninsula or the shops of the West End of London, and each is attractive because it is a unique landscape and environment. In other countries people seek the stimulus of new landscapes by visiting the Grand Canyon, the Pyramids, Ayres Rock or Spanish beaches. Within Britain the diversity of landscape is considerable, although compared with the rest of the world there are few environmental extremes of aridity or wetness, wealth or poverty, mountain or plain; or indeed of any of the other facets of the environment which together combine to make up the landscape.

Sometimes the landscape we see is confined to a small area; sometimes it is spacious. But whether it is of suburban garden, barren mountains, residential tower blocks, or vast fields of wheat, the landscape is the combined result of the activities of present and past societies changing and modifying the natural environment created by physical processes. The core of geography is understanding how the physical and social environments interact to create different landscapes in different places. This is the interplay between ecological and spatial systems that Haggett refers to in the quotation at the start of this book.

Geographers describe the landscape, then attempt to analyse it by unravelling the relationship of all the different landscape agents. These forces may be as diverse as wind, government, glaciers, farmers, rivers or shoppers. Geographers try to explain how and why places are different, and may go on to suggest how the landscape will evolve in the future and even how changes could be controlled. The job of the geographer, therefore, is to describe, analyse and explain the landscapes around us and then to consider how these landscapes will evolve.

Perhaps not surprisingly, few geographers claim to be able to do the whole job and there is considerable division of labour. The major factors responsible for landscape variation have each generated sub-disciplines – branches of geography such as geomorphology, economic geography, political geography, biogeography and historical geography. Alongside these sub-disciplines which are defined by content, there are also sub-disciplines which have grown out of the variety of techniques that geographers use to study landscape: so, for example, we also have regional geography, cartography and spatial analysis.

Landscape evolution

Landscapes not only vary spatially from place to place, as we see on any journey we take, but they also change over time. It is easy to recall how somewhere we know personally, such as a city centre, a suburban housing estate or a particular agricultural area, has changed over the last few years. The city centre may have been redeveloped with a new air-conditioned shopping centre and office blocks; traffic patterns have probably changed, a traffic-free area created, and so on. In the suburbs a housing estate may have been extended over agricultural land; new bus routes may have been introduced; families will have left and new families arrived; clubs and societies will have been formed and changed the patterns of interaction amongst residents; a new school might have been built. Not all the features of the landscape will have altered, however, only a little bit here and a little bit there. These small changes are likely to have taken place in the last few years and are relatively easy to recall, but others will have taken place over a much longer time scale. The suburban housing estate perhaps did not exist at all thirty years ago, and the landscape may have been one of farmland, perhaps of market gardens and greenhouses where fresh vegetables were grown for a nearby city. The particular agricultural land use might have depended on local soil conditions, micro or local climate, and local surface features, the topography. Even longer ago the landscape might have been different again, perhaps part of some cooperatively organised open field system producing most of the food for a small village. Earlier still it probably would have been woodland, and before that 'tundra' or even ice covered. Some attributes of the garden soil of present day British suburban estates can be traced back to tundra conditions 10 000 years ago. Small areas of open space on the same residential estate may have previously been common land for many centuries. The boundary of the estate may be related to land ownership patterns dating back many centuries, and some estate roads are probably built along old farm tracks. The imprint of former landscapes can often be seen in the present day landscape as essentially small features of earlier landscapes showing through the imprint of present day society.

With each occupation and use of an area the landscape is changed, yet parts of the earlier landscapes show through. The process is like patching the paintwork on a house. If each year we go around the house and repair the paintwork, and if each year we use a different colour, after several years small areas of original paint would still be seen, but much of it would have been covered by work in later years. Even some of the repairs would themselves have been repainted and covered over with a new colour. Landscapes that we see today are like this multi-coloured house, with some features relating to occupations by different, earlier, social, economic and political groups of people and some features relating to earlier physical conditions – for example, to previous river channels, vegetation distributions or past climates. Old and new landscape features exist side by side in present landscapes. Exeter Cathedral in South West England, dating from about 1275, exists, functions and contributes to the urban landscape of Exeter within barely 100 metres of the Guildhall shopping centre, which is a landscape feature completely typical of the economy of post-industrial, late twentieth century Britain. In Tallahassee, Florida, remains of Indian settlements 400 years old are conserved within rapidly expanding residential suburbia. In Bangkok, Thailand, modern offices exist alongside ancient temples. Throughout the world features of earlier landscapes show through as major elements in the present day landscape. But some features of these earlier landscapes have disappeared altogether and we may only know of their former existence by residual features. Many of the deserted villages

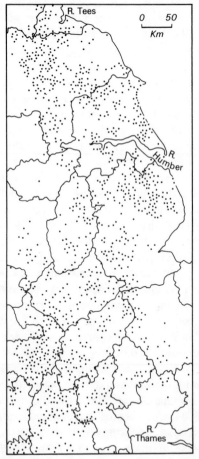

Fig. 1.1 The locations of deserted villages in Eastern England.

(Fig. 1.1) in Eastern England only show in the present landscape as a slight unevenness in pastureland; some others we only know about from written historical sources. Evidence of their existence has become less and less since they were deserted, mostly in the fifteenth century, although at their peak these villages and the society based on them would have been an important influence on the landscape of the time.

Man–land relationships and landscape systems

When considered in this way, every landscape from the bustling city centre to a remote mountain area may appear to be a chaotic mixture of features jumbled together in apparently limitless and structureless combinations. But order does exist. Some associations of features are common; others never occur. This order comes about first through the modifications man makes to the natural landscape; and secondly through the relationship between those features of the man-made landscape resulting from a culture (its society, economy, political organisation and so on) and those features resulting from the physical environment. This 'man–land relationship' has caused considerable debate amongst geographers over the last 150 years. One extreme school of thought called *determinism* maintained that the cultural features were determined by physical controls, particularly climatic ones. Man was considered a *passive* agent in the landscape. At the other extreme were the *possibilists* who argued that man is an *active* force; that everywhere it is possible to create one of several human landscapes, and that man chooses which one to create. Both these extreme positions are currently out of favour and most geographers now consider that man and nature act and react on one another in a complex way so that both are active participants in the development of landscape. Chain reactions are initiated.

Consider, for example, the sequence possible when urban expansion spreads development onto a river flood plain. Floods occur, so a dam is built to control the river flow; farmlands and dwellings are flooded by the lake formed behind the dam and land use around the man-made lake has to change (see Chapter 17 for a discussion on flood prevention). Forestry may take over from grazing if the lake is in an upland area, and in due course the woodland cover promotes changes in soil conditions and new habitats for birds and animals are created. Landscape changes become chained together not only in time but also over distance or space, so that urban growth on the flood plain may be related to agricultural change several hundred kilometres away and several decades later. It then becomes necessary to distinguish clearly between the time and space element. But although the events in time and space are related, there is no rule or law relating the two. Another city expansion may not initiate the same chain of events.

The man–environment relationship may be shown crudely in a simple box diagram (Fig. 1.2). Various cultural and environmental

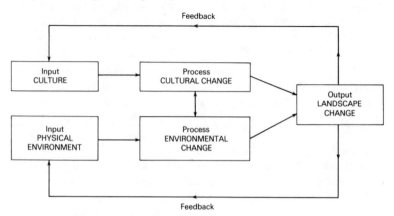

Fig. 1.2 The man-environment relationship results in a system of landscape change

stimuli, or *inputs*, induce some cultural change which itself then initiates environmental change. The results of this change are new environmental and cultural conditions which are called the *outputs*. Within the overall process of change there is *feedback*, in which reactions in the system influence conditions elsewhere in the system as the processes operate through time. Two common types of feedback are distinguished. First is the case where feedback serves to reduce the intensity of the process. Secondly there is the reverse case, where feedback enhances a process and increases the rate of change. The first type is usually termed negative feedback, whilst the second is known as positive feedback. In the legend of the Sorcerer's Apprentice a magic broom is made to carry water, but the apprentice is unable to de-magic the broom and the home becomes flooded. The spell used by the apprentice has no feedback mechanism: hence the disaster. Environmental systems without feedback are similarly likely to be difficult to manage. Within the man–environment relationship the feedback mechanism can in some instances be so strong that it causes a halt in the process. Mismanagement of a landscape, or exploitation of a resource, may result in such pollution that the landscape or resource is no longer capable of being managed or used because it has become useless. The distribution map of deserted villages (Fig. 1.1) relates to conditions at points in space but operating through time. Alternative systems could be described which operated quite quickly but had a wide spatial effect, such as the spread of cattle disease through a grazing region. The example of the chain of events following urban expansion onto a flood plain has both a time and space component. This relationship between time and space is a complex one, but it is important for geographers to unravel it in order to explain the way landscapes change.

This important idea of landscape change being a system with inputs, outputs and feedback is one that provides a framework for geographers. Consider again those recalled changes in the city centre, suburban estate or agricultural area, which were mentioned earlier in this chapter. Various economic, social and political decisions (inputs) would have been made in addition to the selection of an area with suitable physical conditions before a suburban estate (output) could be built. Environmental changes occur which are also outputs from the system. For instance, estate roads and roof surfaces will produce increased water runoff, and wildlife habitats will be radically altered by the influx of human population. Evaluation of the feedback in the system would affect not only expansion of the estate, but also any attempts to conserve specific habitats and future plans for similar estates. The landscape we see around us therefore results from the accumulated interaction of man and environment, with the form of this interaction itself changing due to feedback. Any landscape contains some outputs of the present day 'man–environment system', which generate the process of landscape change operating *now*, but also contains remnant outputs from former processes related to earlier man–environment systems. Using this framework we can begin to get a glimmer of understanding of why the landscape at our journey's end is different from that at the start. We can also begin to see why it is sometimes difficult to decide exactly what in the landscape has changed, or why it has changed.

Changes in the man–environment relationship

The changes in the man–environment relationship in any landscape are most clearly seen in an historical context, with changing evaluations of the environment associated with the development of economic, social and political organisation. In Britain, for example, the period of the Industrial Revolution resulted in a major re-evaluation of the environment and the creation of new man–environment systems. Similarly the transition to post-industrial society taking place in the late twentieth century is having comparable far-reaching effects. In the landscape-related processes associated with iron and steel production these changes are typically seen. When the society and economy were at the stage of using charcoal as a power source for iron smelting, man exploited the oak woods in areas of iron ore bearing rocks. In the Forest of Dean in the South West Midlands and the Weald in South East England, charcoal burning and iron manufacturing were closely interdependent in the late Middle Ages. In the Weald, streams were dammed to form ponds which supplied a water source with sufficient power to work the mechanical bellows and hammers of a furnace and forge. In 1574, records show that thirty-eight forges and thirty-two furnaces were working in the Sussex part of the Weald. With changes in society and the coming of the Industrial Revolution, smelting technology advanced to be capable of using power sources derived from coal. The oak woodlands declined as industrial districts, whilst expansion took place in the coalfield areas such as inland South Wales and West Yorkshire. The greater ease of transport of goods and energy in the twentieth century, together with the need to import more iron ore, resulted in further re-evaluations of man–environment relationships and increased importance was given to coastal and estuary locations. Currently, in post-industrial Britain, another re-evaluation is underway with more complex economic and political considerations now involved and greater international trade in iron products. The result is yet another change in regional patterns of production and a reduction of active locations.

To illustrate such changes Fig. 1.3 shows the distribution of iron, steel and tinplate activity in a small part of South Wales over the last fifty years. The older, scattered plants have been replaced by large, automated strip mills at coastal locations, but even these are now under threat of closure as economic conditions change yet again. A map of 1990 will probably show no places of activity of these

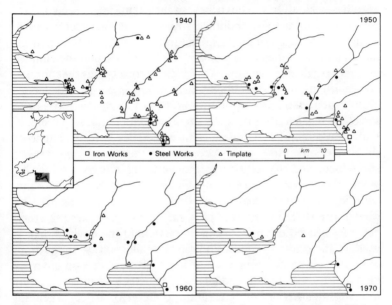

Fig. 1.3 The location of the iron and steel industry works in South Wales. Between 1940 and 1970 fewer but larger works emerged

industries in this area. The industry has changed from being labour intensive, with many workers required in the production process, to being a capital-intensive one for which large amounts of finance are needed to provide high levels of technology. Different economic and social inputs now dominate the pattern of production of iron and steel. What once appeared to be constraints of the physical environment no longer have much importance, and the constraints in the 1980s are of a political nature relating to world patterns of production and demand and national government and EEC policy initiatives.

The many landscape changes which have occurred in Britain can be used to illustrate the man–environment relationship, but the wider range of environments present in other parts of the world is matched by a broader range of cultures and man–environment systems. Consider the variations represented by Chinese merchant communities in Hong Kong, cooperative agriculturalists on the Hungarian Plain, suburban society in Los Angeles, or the subsistence-based New Guinea hill tribes. In each of these cases, economy, religion, social activity, political structure and the man–environment system are quite distinct. The relationship between environment and man depends to a considerable extent upon the cultural values of particular societies. An exploitive society may seek to treat the environment as an endless source of raw material for society's gratification, and little consideration is given to the needs of future societies. In such cases the man–environment relationship is essentially one of conflict. Ultimately such societies have to change or they are doomed to failure. Alternatively there is the approach of coexistence and cooperation, with man living in harmony with his environment. Social and environmental processes act together in a coordinated way and are managed by some political organisation within society. How different physical environments are used then becomes a decision made by the society, and the environment is used for the common good. While such a philosophy sounds attractive, it rarely works in practice. Frequently decisions on environmental use are left to individuals, who may seek to maximise short term personal advantage, so creating pollution for society in general (see Chapter 15).

Landscapes result from a mixture of decisions by individuals and social groups. For example, within the limits set by the physical environment, such as climate and physical features, a farmer decides how to use his various fields. His particular decision will be affected by the society in which he lives and of which he is part. If there is a social or religious taboo against using pig meat, he is unlikely to use his land to rear pigs; if particular crops have to be grown to provide subsistence for the village, then these crops will be grown in preference to others; if there are governmental quotas on the permitted production of a cash or commercial crop, then this will also affect the farmer's decision. The decisions open to the farmer may also be limited by factors such as land ownership. If he

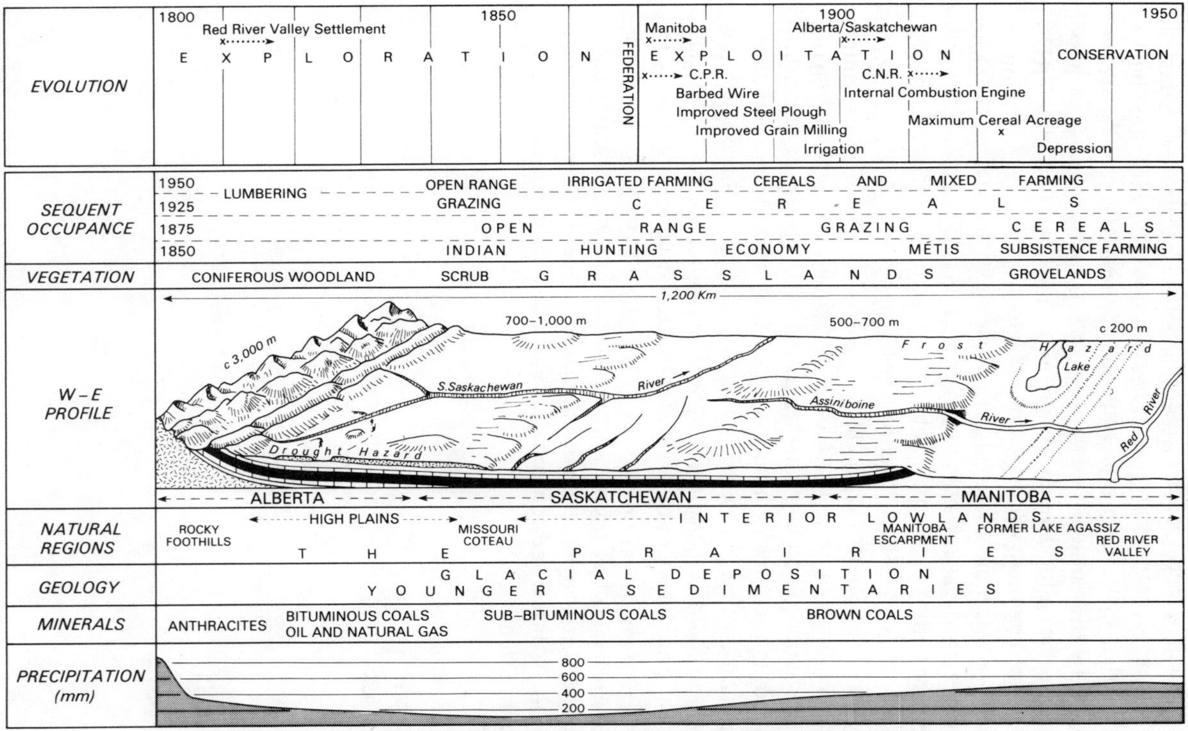

Fig. 1.4 A cross-section of the Canadian prairies showing the associations between man's activities and his environment

owns his land then a greater range of personal possibilities exist than if the land is communally owned by society and the farmer manages the land on instruction from some group in society, or if the land is rented by the farmer. The varied users of urban landscapes have comparable cultural influences on them, and the urban landscape is influenced by these various cultural factors.

The factors which determine land use show the interrelationship between physical and human environments and the man–environment system. An attempt to relate these factors together is shown in Fig. 1.4. A west–east transect is shown through the prairies of Canada. The landscape changes from west to east, but it has also changed through time. The area was originally occupied by Indians with a hunting economy and the pattern of occupancy has changed, but not in the same fashion over the whole area. The exploration, exploitation and occupancy patterns are superimposed upon the physical characteristics of the area. The physical factors underlying land use are the topographic and geological structure of the land, its soil cover and the climate. The human factors relate to the economy, in respect of the returns on investment in alternative uses; to society, in respect of the communal demands of groups of people; and the political and religious organisations which govern land-use management.

2

Resources, Regions, Nodes and Networks

The interdependence between man and the environment is complex. Each has a significant impact on the other, with man using and abusing the environment in several ways. The environment provides a variety of natural resources which are made useful by man's activity, and man also has generated the wealth of human resources. Most natural resources vary greatly in their occurrence, composition, quality and location. Iron bearing rocks, for example, vary in their iron content from a mere trace to over seventy per cent of iron. The particular influence of an iron ore deposit also varies according to the availability of other natural resources, such as coal, and human resources, which in turn depend on many social, political, economic and technological factors. The presence of a usable resource is a major factor, but not the only factor, influencing patterns of land use in a landscape and some natural resources can be used in several ways. The concept of a resource is a complicated one, but it is important in understanding and explaining landscape variety.

Classification of resources

First, it is necessary to distinguish between renewable and non-renewable resources. *Renewable resources* include the living or biotic resources which are used in farming, fishing and forestry and which depend on a steady energy input from the sun. Although this type of resource, such as a forest, may be destroyed by use, the period of regeneration is relatively short. *Non-renewable resources* are those exploited on a once and for all basis, and as they are used their amount is reduced. Examples of non-renewable, or finite, resources are the fossil fuels such as coal, oil and natural gas, and most minerals in the earth. This simplified division of natural resource types is shown in Fig. 2.1.

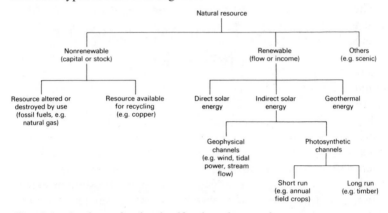

Fig. 2.1 A scheme for the classification of types of resource

Another division can be made between economic resources and social resources, but it is not always easy to distinguish between the two, as for example in forest areas which are also recreation areas. The felling of timber in the forest makes it an economic resource but the recreational use makes it a social resource, the social resource being effectively destroyed by its economic use. Attempts are sometimes made to put a financial value on the social uses of a resource, but usually such attempts are unsuccessful because of the fundamentally different system of values underpinning economic and social resource use. The view of what is a social resource and what is an economic one depends to a considerable extent on the resource appraisal carried out by society. As societies change and new values emerge, so resource appraisal changes. The increased environmental and ecological consciousness of European and American society in the late twentieth century is resulting in radical reassessments of the distinction between social and economic resources. Mineral extraction in an area of natural beauty would have been carried out without any reference to landscape quality a century ago, when economic evaluations usually had precedence over social ones. Since the mid-twentieth century social evaluations have become more important and the social resource of natural beauty in a landscape might be considered more valuable than the economic return of extracting the particular mineral. Many assessments of resources are specific to particular societies and cultures.

The changing nature of resources is also shown by the usual classification into stock, resource and reserve. *Stocks* may be considered as the total amount of a particular feature of the environment. So, five per cent of the earth's crust is iron; or the solar energy received by the earth is 17×10^{13}kw per day. From these stocks, *resources* are drawn. Much of the stock at any one time holds little interest for man because for some reason it is unusable. As technologies and societies change, various items of stock become resources. Alternatively, it is possible for resources to be returned to stock because social and technological change renders some mineral or energy source redundant. Flints for tool making and oakwoods for charcoal burning are examples of this type, and if some conservationists have their way, uranium could revert to stock in the future although it has only recently moved out of this category. *Reserves* are that part of resources which are currently not being used but are quite close to being used. Material is transferred from reserves to resources as current resources become depleted through use or when socio-economic change results in an increased demand being made on current resources. Alternatively demand may falter and resources are consigned to reserves. Considering natural resources in terms of stocks, resources, and reserves stresses the way resources change as society's use for them changes.

Resource assessment

Natural resources are, of themselves, passive and only become active aspects of the landscape when they are exploited and used by man. The presence of coal, iron-rich rock, or a soil of a particular composition has significance for the natural landscape but not, of itself, for the man–environment relationship. The presence of a particular combination of soil and climate will allow the development of a specific natural vegetation, but the soil and climate are part of the stock of resources until the potential is realised by man and they are used for agriculture. Most geographers argue that the word 'resource' relates not simply to a thing but to a function which the object may perform – the function of satisfying some aspect of man's wants. Resource assessment or the study of resource function is a complex topic. There is no deterministic relationship between the presence of large quantities of reserves and their use, or between the presence of natural resources and the level of economic development. The lack or abundance of a natural resource is not a determining factor in economic development. The critical factor is the assessment of the resource in the light of technological, social and economic constraints on resource use.

At least four conditions have to be met before a resource is used and it comes to have an active effect on the landscape:

1   The physical presence and human awareness of the resource.
2   A technology available to extract and use the resources.
3   An economy, with suitable managerial skills, which would benefit from use of the resource.
4   A culture and society in which the resource is an acceptable commodity and in which its use does not offend social taboos.

The presence of all four preconditions will usually mean resource exploitation. This may be the creation of electricity at hydro-electric power plants, iron ore mining, salmon fishing or the provision of campsites in wilderness areas. Resource use is considered in more detail in Chapter 3.

Geographers study resource use and its landscape effects in order to understand the man–environment relationship. In describing these resources and landscapes, three terms – *region*, *node* and *network* – are often used, both in respect of man's activities and of processes in the physical environment.

Types of region

Three major types of region are recognised by geographers. The *formal region* was defined by Professor Hartshorne, an American geographer, as 'an area of specific location which is in some way distinctive from other areas and which extends as far as that distinction extends'. It is useful to be able to talk about the Lake District or the Congo Basin as regions which have location and area and around which we could draw a boundary. These regions have features of homogeneity which make the region, as a whole, distinct from surrounding regions. These formal regions can differ very considerably in size; for example, the European Economic Community (EEC) is towards one extreme and Mayfair, in central London, or Manhattan, in New York, is towards the other. The boundaries of regions may change over time, either with an instant contraction or expansion – as, for example, when Britain, Denmark and Ireland joined the EEC – or more gradually, as in the case of the corn belt of the USA, where evolving agricultural practices over the last 150 years have caused extension and contraction of the boundary of this region of distinctive agriculture. We can divide up the world into formal regions to describe the broad pattern of landscape types, so that we may have world rainfall regions, world soil regions, world agriculture regions and so on (see Chapter 7 for some examples). At such a large scale the mapped boundary lines may be rather inaccurate, but comparison of regional maps allows broad descriptions of world landscape to be made. Formal regions may be defined in terms of a single criterion, for example membership or non-membership of the EEC; or several criteria may be used and combined, perhaps in some mathematical way, to define multi-category regions, as for example in climatic classifications or the study of regions such as the central business district (CBD) within a city.

The second type of region is the *functional region*. This type of region depends for its definition, not on sameness, as in the formal region, but on the idea of dependence. The region is the area which depends in some way on a place within it. A common type of functional region is the area around a city or town. People in the surrounding area visit the town to shop and possibly to work; road traffic flows within the region focus on the town. The general influence of the town spreads out over a functional region. Again, functional regions vary considerably in size and their boundaries can change over time in similar fashion to the formal region. Functional regions allow us to describe some of the interdependence in the landscape, and in some cases may be considered as systems in their own right. The region around a city usually has inputs and outputs with other surrounding regions and within the city region there is spatial interaction amongst its different parts. Flows occur through the region and spatial feedback is present.

Relationships in the city region also change over time, with the results of present activity affecting future events and so giving a sort of temporal feedback.

Thirdly there are *programming regions*, which are defined in a utilitarian fashion to allow man to manage the environment in some way. The creation of economic planning regions within countries is a regional division of this type. The boundaries of the regions may

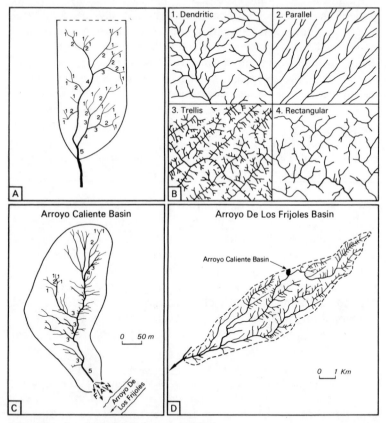

Fig. 2.2 A A method of stream ordering
   B Major types of stream pattern
   C A small fifth-order drainage basin near Santa Fe, New Mexico, USA
   D The larger drainage basin containing the small basin shown in diagram C

be relatively arbitrary to all but the administrators involved and may run counter to, or in some cases be identical with, regions defined in other ways.

This threefold division into types of region is not in practice clear cut. Programming regions can coincide with formal regions and may in some cases be similar to functional regions. In a planning region, for example, there is likely to be a central point of administration, with decisions flowing out from this centre and reactions flowing back to the centre similar to the working of a functional region. In any regional division of landscape the differences in scale are important. Smaller regions are often enclosed or *nested* in larger ones. Within a drainage basin region, the land area drained by a river and its tributaries, it is possible to consider the basin either as a single region, or a small number of regions, each of which is the drainage basin of a major tributary, or as a larger number of small regions, each associated with a tributary of a tributary, and so on down to a very large number of very small regions each of which is the drainage basin of an individual stream. Streams and rivers can be ordered and numbered: the smallest unbranched streams are ordered as 1; the channels formed by junctions of two first-order channels are order 2; those formed by the junction of two second-order channels are order 3 and so on (see Fig. 2.2A). In similar fashion drainage basin regions can be ordered with the smallest (first-order region) nested in second order, and so on (see Fig. 2.2C, D).

The physical landscape as a whole can be described in terms of such a sequence. The smallest landscape unit in such a scheme is an area with a homogeneous complex of physical conditions including altitude, degree of relief, rock type, soil, vegetation and drainage and may be a flat valley floor, a flat plateau surface, a steep slope, etc. Such a unit is called a *terrain facet*, or sometimes a biotope or ecotope. A group of these facets may occur in sequence commonly enough for them to be a region in their own right. The combination of valley floor, slopes and interfluve, for example, is a common association of facets. Such combinations are termed a *stow*. Stows may be combined to form *tracts*. An example of a tract is the Lake District. In turn these tracts are part of fourth-order features such as Highland Britain – that area of Britain west of the Tees-Exe line. The combination of these regions forms *provinces* or *realms* of sub-continental proportions such as North West Europe, and again these may be associated into world climatic-vegetation regions called *biomes*. The smaller units in this scheme are combined to form successively larger regional units. This nesting is also found within functional regions such as the nesting of the service areas of settlements of different size. The area of dependence, the functional region of a settlement, is related to its size and to the range of functions it offers to its region. Large cities provide many functions which serve people from a wide region, but within this region are towns offering fewer attractions and commanding a smaller 'tributary' region. Within this smaller region are villages which offer fewer services and have an even smaller functional area (see Chapter 11). The functional regions of each size of settlement are nested in the functional region of successively larger towns. The real world is a complex mixture of regions of different kinds and of different scales but regions are only one component within the landscape.

Patterns of nodes

Nodes are a second component feature of the landscape. They are points of concentrated activity and as such are basic elements of the landscape. Examples of nodes are cities, the epicentres of earthquakes, waterfalls, carparks, even schools. In all cases, these nodes act as foci of activity and interaction with the surrounding area. Clearly nodes are an integral part of functional regions, but also they exist in many formal regions. Within broad resource regions nodes provide the places of active resource exploitation. In a coalfield region one group of nodes might be the mine sites acting as foci for the journey to work of miners and the points from which coal is distributed by rail and truck. These nodes would also be the centres for accident services, major points of investment and perhaps concentrations of housing.

Within a region we can think of nodes as being points scattered throughout the landscape or, as it is often termed, through the regional space. The nodes are distributed in regional space in different ways, making different patterns of dots on the regional map. There are three commonly recognised types of distribution or pattern of nodes – regular, random and clustered (see Fig. 2.3) –

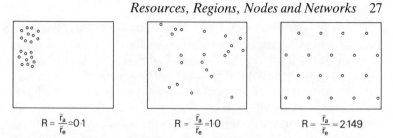

$$R = \frac{\bar{r}_a}{\bar{r}_e} \simeq 0.1 \qquad R = \frac{\bar{r}_a}{\bar{r}_e} = 1.0 \qquad R = \frac{\bar{r}_a}{\bar{r}_e} = 2.149$$

Fig. 2.3 Clustered, random and regular patterns of dots

although the actual pattern of dots can vary anywhere between the extremes of regularity and total clustering. In a regular distribution we imagine nodes at the apexes of many equilateral triangles, so that within the overall region the nodes are equally spaced. The other extreme is total clustering, when all the nodes in a region congregate at or are close to one place. Between these two extremes is the pattern where the nodes are scattered in a haphazard, or random, fashion throughout the region.

It is sometimes difficult to decide the type of distribution just by looking at a pattern of dots or nodes. It may be necessary to take measurements and compare these with a key which shows where the distribution lies along the scale of measurement from clustered to regular. The technique is usually called *nearest-neighbour analysis*. The distance from each node to its nearest neighbour is measured. If there are twenty nodes, there will be twenty nearest-neighbour distances. Sometimes it is necessary to measure a distance to a node outside the immediate region of interest if the nearest node is across a regional boundary. The average of these nearest-neighbour distance values is then found. On Fig. 2.3 this is written as $\bar{r}_a$. The density is then calculated simply as the number of nodes per square unit. For example, if the 20 nodes lie in a rectangular area 3 km by 2 km, then density is 20 divided by 6, which equals 3.3 nodes per square kilometre. The next stage is to calculate the square root of the density measure and then halve it. So the square root of 3.3. is 1.8, and half of it is 0.9. This value ($\bar{r}_e$ in Fig. 2.3) is what the average nearest-neighbour distance would be if the distribution were a random one.

The final stage of analysis is to compare in a ratio the two numbers we have calculated as the average actual neighbour distance and the density related random distance. The ratio of these numbers ($\bar{r}_a : \bar{r}_e$)

will have a score between 0.0 and 2.149. The closer the value is to 0.0, the more clustered the distribution will be; the closer the value is to 2.149, the more regular the distribution will be. If the value is close to 1.0, then the pattern of nodes will be a random one (see Fig. 2.3). By using this measurement method it is possible to describe in an accurate way a pattern in the landscape, such as the location of tors on Dartmoor, the pattern of villages in part of Lowland Britain, or (as in Fig. 1.1) the pattern of deserted villages in Eastern England.

The pattern of nodes in the landscape changes as the landscape itself changes. New features appear in a landscape and may spread through it. These features can often be plotted as nodes on a map of the region and the changing pattern of these nodes shows how the new feature spreads through the region. The form of the distribution may well change as the feature invades the region and *diffuses* through it (see Chapter 14). Many changes in the landscape occur in this way as some new resource use enters a region and spreads or diffuses through it, or some ecological change enters and spreads through an ecosystem.

Networks in the landscape

The third component of the landscape is the *network*. One type of network provides the pathways along which interactions in the landscape occur, and such a network links together nodes. A stream, rail or telephone system is a *flow network* of this type. Alternatively the boundaries of regions may be considered as networks, as for example in the pattern of field boundaries, or in the pattern of stone polygons formed by the natural sorting of stones in areas of permanently frozen ground. Boundary intersections may then be considered as a particular type of node. Whether a flow or *boundary network*, the network pattern may be represented in a diagram by a series of lines. Networks are important in the analysis of landscape because most of the movement of resources, of people, goods and even ideas, takes place or can be represented as taking place on flow networks. Consequently networks are vital to the study of processes of landscape change and resource exploitation. To take a simple example of a hydro-electric power scheme: whilst activity takes place at the power plant (node), this activity is not integrated into the region until a network of power transmission lines is built. The activity also depends on the flow in the river network which concentrates water energy at the exploited node on the river.

It is usual to distinguish two basic patterns of network. First is the *circuit* network, in which it is possible to start at a point and return to it without retracing a path. A circuit network pattern does not mean that flows actually do occur around the circuit, only that the level of connection is such that circular flow could take place. Examples of circuit networks are a road system in a developed country or the pattern of channels in a saltmarsh. Secondly there is the *tree* network which has no circuits, but, as the name suggests, is a major link to which branches join. An example of this type of network is a river system (see Fig. 2.2A), or the journey patterns of visitors to hospital or children to school. The important distinctions between these two network patterns relate to the *centrality* of particular points or nodes on the network and the *connectivity* of the total network.

The centrality of a node on the network is the ease with which it is possible for interaction to occur between one node and other nodes. In the tree network there is a very large range of centralities associated with different nodes. A node at the tip of a branch in the tree network has a low centrality but the key node – for example, the school or hospital – has a very high centrality. In circuit networks the variation of centrality values amongst the nodes is likely to be smaller as it is easier to interact across the network. As general interaction levels are potentially higher, so *connectivity* is higher. In a circuit it is possible to connect two nodes by more than one route, and in networks with a high degree of connectivity several alternative routes may be available. The passenger network of British Rail, for example, allows journeys to be made between major towns by several routings, but passengers usually choose one route because it is shortest or quickest, or cheapest, or some combination of these. Sometimes, however, a passenger might wish to string together visits to several major towns, ultimately returning to the starting point. This is possible on a highly connected network without having to backtrack. Thus a journey around England from London—Exeter—Bristol—Birmingham—Manchester—London is possible because of the high level of connectivity in the rail

network. If, for example, no link existed between Bristol and
Birmingham a passenger might have to return to London in order to
travel out again to Birmingham. Fewer links means lower connectivity. Connectivity on the British Rail network is quite high. In
countries with a less well-developed rail network connectivity is
lower and tree networks are more common. A complicated journey
around several towns on a tree network would probably mean
back-tracking to a key node of high centrality and then travelling
out again. In these networks connectivity can sometimes be increased dramatically by the addition of a single connection. When
long distance travel used to depend on river transport, for example,
then portages often had to be used with, the traveller carrying his
canoe, walking, or riding across land to the next river system or tree
network. Later, canals were often built at such portages to provide
an extra link and so extend a network, or in some cases turn two
adjacent tree networks into a circuit network. The same principle is
often used in extending the rail networks in developing countries,
allowing increased connectivity in the network with the minimum of
capital investment.

There are many ways of measuring networks and assessing
centrality and connectivity. The production of distance tables of the
type used in road atlases showing the road (network) distance of
each place (node) from every other place (node) is a simple way of
measuring a network. Additional measurement methods have to be
used to find out more about the routes or connectivity. Brief
mention has already been made of one method of ordering stream
channels and their associated drainage basin regions (see p. 24).
This sort of analysis allows the study of the relationship between the
numbers of streams of different orders within a single high-order
drainage basin, between similar drainage basins, perhaps on different rock types or in different climatic regions, and also between
streams of different pattern, for example, amongst the four main
types of drainage pattern – dendritic, parallel, trellis and rectangular (see Fig. 2.2B). A variety of analyses are possible using this basic
method and considering the ratios of the numbers, and lengths, of
streams of different orders.

An alternative way of measuring networks uses the same method
for tree and circuit networks, considering the ratio between the
number of links in the network and the number of nodes (or

junctions). The simple ratio index of links:nodes, sometimes called the ß index, has the useful attribute that it describes a network irrespective of its size. The greater the index value, the higher the connectivity in the network, with values below 1.0 identifying tree networks and values of 1.0 and above identifying circuit networks. The usefulness of such a measure is that it allows comparison amongst networks. When several railway networks are compared, for example, we find that Ghana, Bolivia and Sri Lanka have values of around 0.9 – signifying low levels of connectivity – so that there is a high likelihood of backtracking on multi-node journeys. By comparison, France has a value of over 1.4, indicating a complicated, highly-connected network. If the values for many countries are considered, then four groups emerge. First are the countries, examples of which have been mentioned, with only a partially-connected network and a ß value around 0.9. Next are a group of countries typified by Turkey, Algeria, Iraq and Thailand, which have ß values of about 1.0, indicating a simple circuit in the rail network. Thirdly there are the reasonably well interconnected networks such as in Mexico, Rumania and Yugoslavia, with ß values between 1.15 and 1.3. Finally there are a group of countries with an advanced level of connectivity and ß values of around 1.4. Most West European countries fall into this final group. The degree to which there is connectivity in a country's transport network is one measure of economic development (see Chapter 8), with more complex economies requiring more highly-connected transport networks.

Relationships of regions, nodes and networks

The regions, nodes and networks in a real landscape are all inter-related and their individual treatment in the previous few pages perhaps overstresses their individuality. Networks can indicate the relationships amongst nodes or show the edges of regions. Regions contain patterns of networks and nodes. Nodes provide central points for networks and regions. Together the three 'building blocks' create a landscape which functions through the interdependence and interaction of many types of node and network. Geographers, in analysing the landscape, try to understand what processes bind together these components to make a landscape work. From the geographers' standpoint, energy and information are the 'cement' of the landscape and environment. Solar radiation provides the ultimate motive force for present biological and physical landscape activity, and many of the fossil energy sources used by man represent the trapping of solar energy many million years ago. In the human landscape, the economy and society of man depends upon information in its widest sense. Cities exist because of the needs of people, through their economy and society, to pass information to each other. A region of commercial farming depends not only on energy inputs but also on a mass of information on agricultural practices, markets, prices, farmers' behaviour, and so on. In order to understand how the component 'building blocks' become 'cemented' together, we need to return to the idea of systems within the landscape. We need not only to describe what we can see in the landscape, but also to try and look at how the energy and information based processes place patterns of regions, networks and nodes in a constant state of change.

3

Landscape Processes and Environmental Systems

Some ideas of landscape differences were introduced in Chapter 1, together with the concept of a landscape system which varies through geographical space and through time. In Chapter 2 we looked at resources in one landscape and at ways of studying landscape by describing its component regions, nodes and networks. In this chapter we go on to consider changes in resources and the consequential ways that landscapes change.

To begin to try to analyse the changes we need to measure their speed and describe precisely what changes are taking place. Resources are used at different rates, at different places, at different times, resulting in one set of contrasts in regional systems. Major iron deposits in Brazil and Western Australia have been exploited differently, not only because of the differences in the physical composition of the ore in the two regions, but also because of other factors such as differences in ownership, differences in political attitudes towards ore exploitation, and differences in the cost structure of exploitation, which are themselves related to the particular location of the ore body, its ease of working and the available markets. Generally the stock of a resource, in this example, iron ore, consists of large quantities of low grade, poor quality reserve and small quantities of high quality resource. This is sometimes termed the '*resource pyramid*' with quality declining towards the pyramid base. The parts of the resource which are exploited are usually as close to the apex as possible, but technological invention or some political upheaval can result in the economic exploitation of lower grade resources and a transfer is then made from reserve to active resource. Generally, despite non-renewable resources being used constantly for many centuries, there is little overall evidence of increasing resource scarcity. Increases in the relative costs of natural resource commodities have been avoided for four reasons which might be called the rules of resource use:

1   As higher grade sources are exhausted, lower grade resources are found in greater abundance – the base of the resource pyramid is expanding.
2   As a particular extractive resource becomes more scarce, the rate of increase in its price tends to be offset by substitution of other resources.
3   Increases in price stimulate greater search activity to find new deposits and provide incentives for recycling, so reducing pressures on virgin resources.
4   Technological change is directed toward reducing the costs of providing natural resource commodities, both through reduced extraction costs and by encouraging transfers from stock and reserves to active resources.

The majority of resource development patterns, over the last 100 years, broadly fit this pattern. Implicit in these rules is the idea of transfers taking place between stock and active resource. When such a transfer is made, usually it means that resources are exploited in a new location and there is a sudden landscape change and a rapid change in the man–environment relationship. More commonly resource exploitation involves a gradual change in the landscape as activity takes place.

The speed of landscape change

A gradual change interspersed with brief periods of rapid change, a sort of 'stop-go' sequence, characterises many patterns of resource use and in turn broader landscape change (see Chapter 13). The draining of the Fenland area of Britain provides a good example of such relationships. Initial draining took place quite quickly and separated two periods when landscape change was more gradual. Before the early years of the seventeenth century, attempts to drain the Fenland area were limited and piecemeal and made little impression on the overall landscape. In a thirty-year period from about 1625, however, both political incentives and technological change allowed rapid development of a unified drainage scheme with corresponding major landscape change. Cultivation was introduced on lands which never had been ploughed before and the agricultural area was expanded. After this surge of activity the landscape still continued to change, but more gradually, due to the lowering of the drained peat surface as the soil and peat dried out, and also as the dry top surface washed away. Windmill pumps, and later other forms of pumping, had to be introduced to lift the drain water from field and small drains into the main drainage channels. The rapid landscape change through drainage changed agricultural, economic, social, hydrological and many other features of the environment, and in fact generated feedback which created further change, although of a more gradual nature.

A second very different example shows a rapid landscape change taking place today. Large quantities of the workable coal deposits of the USA lie close to the surface or outcrop along hillsides and are extracted by strip-mining methods. In this method earth-moving equipment removes the spoil material which overlies the coal and then the coal is dug out with power shovels. There are two main types of strip-mining – area strip-mining and contour strip-mining – and in each case ridges of spoil are piled up as the coal deposits are uncovered. As mining takes place the landscape is completely changed from farm or woodland to derelict spoil heaps. Over 1 600 000 hectares of land have been affected in this way in the USA, and whilst most of this has been in the Appalachian region of the east of the USA, there is increasing use of these methods in the massive mines being developed to exploit the extensive coalfields in the western states of Montana, North Dakota, Wyoming, Utah, Colorado, Arizona and New Mexico. Coal output from these areas is increasing rapidly, and in Wyoming and Montana the seams of coal are frequently over 20 m thick. Five major strip mines in Montana produce, in the early 1980s, over 50 million tonnes of coal per year. The landscape of spoil ridges which is left after coal exploitation in Appalachia is susceptible to rapid change by physical processes, particularly after heavy rain. In the drier western states the natural vegetation of the area is shrub and grassland, which grows and changes very slowly. The rapid change caused by coal exploitation means that the natural vegetation is destroyed and the spoil in the ridges is not able to store the water necessary to allow vegetation to grow again unless it is carefully, and expensively, regraded. Even when the spoil is restored by the addition of top soil, the regeneration of vegetation is very slow. Once mining activities and resource extraction begins, landscape change is very rapid during the period of mineral exploitation, and becomes more gradual as activity patterns change. Strip-mining can result in a complete transformation of the landscape in only a few years.

There are many other examples which could be described to show how the speed of landscape change is not constant, but rather bursts of rapid activity alternate with long periods of gradual evolution. One way of describing this difference in speed is by considering the changing energy input into the landscape system. In a landscape where man is absent, a steady input of solar and gravitational energy creates gradual change in the vegetation and form of the land. The input of solar energy is uneven over the world (see Chapter 8) but the transition from one area to another is a gradual one. A sudden input of natural energy, say, from a volcanic eruption or a tropical cyclone may cause rapid change in the landscape and can cause widespread devastation. Alternatively, in a landscape where man is generating energy and applying it to the landscape, its application is both uneven and apparently haphazard. The social and economic form of energy which man uses is not only the strength of his arm but, more importantly, money or capital. A sudden injection of capital serves to pump economic energy into the landscape, as in the draining of the Fens and the strip-mining for coal. The apparently haphazard nature of the application of human energy is due to the differences in the amount of information man has about his environment and the resources it contains. Variations in energy and information input into the landscape are responsible for the different speeds of landscape change. Geographers are concerned not only with describing change in resource use and landscape but also in trying to understand these changes by viewing them as the outputs of various environmental systems.

Four major environmental systems

The whole earth may be considered as one huge system, but it is more useful for the geographer to recognise four major environmental systems and to study the process of landscape change with reference to each of these systems. The processes within the systems usually are active at or near the earth's surface. The four major environmental systems are the atmosphere, the hydrosphere (water in its various states), the geosphere (earth's crust) and the biosphere (life in its various forms, including man).

The atmosphere

The atmosphere consists of the envelope of gases which surrounds the earth. Although it extends several hundred kilometres into space, most of it lies within 10 km of the earth's surface, or below the height of the highest mountains. Within the atmosphere there are many processes which affect our environment and some of these processes may themselves be considered subsystems of the main atmospheric system. The weather and the patterns and regularities which are termed climate, consist of several subsystems which although linked to other subsystems are internally consistent. A thunderstorm, for example, may be viewed as a subsystem and consists of a series of small rising air cells, or bubbles. As each rises, so surrounding air is brought in from around the thunderstorm cloud. Fig. 3.1 shows the generalised patterns of air movement in a thunderstorm. When the rising air reaches high levels (usually between 6 and 12 km) the rate of ascent decreases and the cloud top is dragged downwind to produce the characteristic anvil-shaped top to the thundercloud. Ice particles generated at the top of the cloud fall through the cloud and help rain droplets to form. These may get caught up in the rising cells of air and pass up through the cloud, falling down and passing up the cloud perhaps several times and getting bigger all the time. Finally, precipitation occurs, either as rain, or if the rain droplets have been frozen during their movements within the cloud, as hail. Precipitation falls from the cloud when the upward movement of air loses some energy. The thunderstorm forms a good example of a subsystem within the atmosphere as it requires an energy input to start the air cells rising, but then internal changes and processes occur which make the subsystem work. The initial energy necessary to cause air cells to rise usually comes from differential warming by solar radiation, causing a particular mass of hot air. This is most common in tropical areas where solar radiation is most intense and where thunderstorms

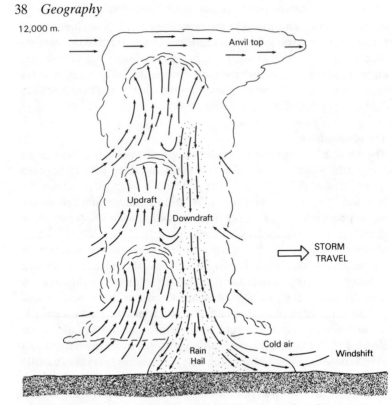

Fig. 3.1 Generalised processes in a thunderstorm subsystem

occur all year. In temperate latitudes they are normally a summer phenomenon, and they rarely occur in arctic latitudes.

The thunderstorm is a clear-cut subsystem with obvious environmental effects. The atmospheric system comprises many subsystems, each of which may be studied individually, but it is also interesting to consider what happens at the junction of subsystems, such as along the boundaries between contrasting air masses. Where warm air moves to displace colder air at the ground a *warm front* develops. The less dense warm air moves up and over the denser cold air and cloud forms, frequently with precipitation, which is often rain. As the actual front passes, the temperature rises, winds shift, precipitation stops and cloudiness decreases. Alternatively, if cold air advances on warm air, then it moves under the warm air and forces the warm air to rise and a *cold front* develops. If the cold air is moving rapidly this uplift given to the warm air may be considerable and cause gusty winds, intense precipitation, and may even generate the energy sufficient to initiate a thunderstorm. The passage of the cold front causes temperatures to drop, winds to shift, a fall in relative humidity and an increase in atmospheric pressure. The boundary between the air masses, between the two subsystems, provides a belt of intense weather activity. These frontal boundaries may separate large air masses (see Chapter 7), run for several thousand kilometres and be associated with features in the upper atmosphere. Alternatively they may be small and localised, perhaps only a two or three kilometre boundary associated with a local sea breeze front. More detailed descriptions of atmospheric subsystems are provided in the books listed in the further reading at the end of the book. Many of the subsystems in the atmosphere are imperfectly understood although considerable progress has been made in recent years in studying the energy transfers taking place within various subsystems. Attempts have even been made to modify some processes in particular subsystems in order to modify weather. Rainmaking, fog clearing and the reduction in hailstorm frequency have all been attempted, but such attempts probably are premature and potentially dangerous when so little is understood about processes within atmospheric subsystems.

The hydrosphere
Another major earth system is the hydrosphere, which consists of water in its many forms in the air, on the ground and below the surface. This system is characterised by a cycle of activity which allows water to circulate through the atmosphere, biosphere and geosphere. The *hydrological* cycle consists of a complex set of processes which cause changes in the state of water to ice or vapour, and which result in its circulation around the earth. The oceans form the principal reservoir of the earth's water (see Table 3.1) and each year an estimated 344 000 km³ of water is evaporated from them and a further 56 000 km³ is accounted for by evaporation from lakes, rivers, soils and vegetation and from transpiration by vegetation. The oceans receive back this 400 000 km³ in the form of rain, with an estimated 96 000 km³ falling on the continents and 34 000 km³ falling

Table 3.1   Estimated distribution of the world's water (cubic km)

	World	Per cent
Water on land		
Freshwater lakes	125 000	0.009
Rivers (at any one time)	1 300	–
Glaciers	200 000	0.015
Ice-caps	28 800 000	2.1
Saline lakes	100 000	0.007
Water in the land		
Soil moisture	65 000	0.005
Groundwater		
Within 1000 metres	4 000 000	0.29
In the next 1000 metres	4 000 000	0.29
Atmosphere	13 000	0.001
World oceans	1 320 000 000	97.0
Total (in round figures)	1 360 000 000	100.0
Annual evaporation		
From land	56 000	0.004
From oceans	344 000	0.025
Total	400 000	0.029
Annual precipitation		
On land	96 000	0.007
On oceans	304 000	0.022
Total	400 000	0.029
Runoff of all rivers		
Per year		
(water yield)	40 000	0.003

directly back into the oceans. From these figures we see that 56 000 km³ is evaporated from the continents, but 96 000 km³ is received in rain. The balancing 40 000 km³ is returned to the oceans by runoff, most obviously through rivers and glaciers discharging into the oceans, but also through seepage and the passage underground of water. Some rain falling on land infiltrates into the soil, whilst the remainder runs off directly into streams and rivers. The amount passing into the ground varies particularly with soil texture, soil moisture content and land use. Sandy or gravelly soils do not swell when they get wet; the gaps, or pores, in the soil structure are relatively large and so the soil has a high ability to accept large amounts of water and *infiltration capacity* is high. In contrast, clay soils have small pores and the soils swell when wet, so infiltration capacity is low. Infiltration capacity will also vary with the amount of water already in the soil. Vegetation cover is also important, with natural forest having a high infiltration capacity because the vegetation and fallen foliage protect the soil from damage by falling raindrops and soil passages remain open. As man has cleared the forest or vegetation, as with the strip-mining of coal in the USA, the infiltration capacity of the land is decreased, runoff is increased and sometimes this has meant that new streams have been formed which have caused soil erosion and gulleying.

Water in the ground infiltrates the soil and gravity causes it to percolate downwards, through an intermediate zone which is too deep to be reached by plant roots, to the water table. Water from this saturated ground level can be extracted by wells dug down through the water table, or it may come to the surface naturally at a point where the ground surface is lower than the water table. This saturated layer of ground needs continuous recharging as the water is steadily being moved out both through the activities of man and by the natural drainage. If recharging does not occur then the water table falls and wells begin to run dry. There may be a seasonal rise and fall in the water table corresponding with wet and dry seasons, but there may also be longer term fluctuations in the water table level due to broad changes in climate: land use, such as when forest is steadily cleared from an area, or, perhaps even more importantly, as man extracts more and more water from the ground reserve. When man has extracted the water and used it, he usually returns it to rivers or back into the soil, so that although his activities are important locally, he barely interrupts the hydrologic cycle on the world scale, as shown in Table 3.1. The water cycle in the hydrosphere has many such subsystems within it, several of which intercept water at some stage of the cycle, use it and then return it to the basic cycle.

The geosphere

The third major earth system is the geosphere, and most of the processes in which geographers are interested occur in the outer 8 km of the earth. There is now widespread acceptance of a theory of global plate tectonics which helps to explain the present day position of the continents and oceans in the geosphere. It is assumed that the outer part of the earth, the *lithosphere*, is made of several enormous brittle sections which effectively float on a softer earth interior (*asthenosphere*) (see also Chapter 15). These sections or plates are constantly moving, and where they move against each other large amounts of energy are transferred. This energy transfer can cause movements such as mountain building, earthquakes or seismic activity, and volcanic activity (see Chapter 10).

Originally six major plates were identified, but it is now recognised that there must be at least twenty smaller plates which move independently of each other. The edges of the plates are mapped as lines on Fig. 3.2. Each plate, which consists of both crust and upper mantle, can carry both ocean and continent and there is no simple relationship between continental positions and plate positions. As the plates move, their edges come into contact in different ways and

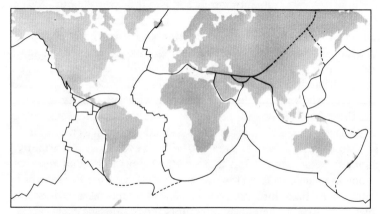

Fig. 3.2 The simplified pattern of tectonic plates

three types of margin are recognised. These are termed *constructive*, *destructive* and *conservative* margins and are shown in Fig. 3.3. At constructive margins new material rises from beneath the junction of two plates and spreads out as it reaches the lithosphere. The large scale crust and mantle movement, or tectonic activity, along a North–South line in the mid Atlantic Ridge, marks a constructive margin between plates. At destructive margins one plate passes down and under the second plate and material is returned to the asthenosphere. This process of absorption is called *subduction*. The destructive edge of plates is often marked by the deep oceanic trenches which are areas of considerable earthquake activity. World distribution maps of the occurrence of earthquakes show a marked zone of activity in the North Pacific, through Japan and the Philippines and eastwards through New Guinea and the islands of the South Pacific. This line represents the destructive margin of the Pacific plate and probably also of several smaller ones. At a conservative margin two plates slide past one another without any gain or loss of plate. Such a transverse movement also generates major earthquakes and the western edge of the American plate and eastern edge of the Pacific plate make up a conservative margin

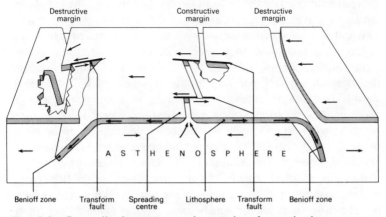

Fig. 3.3 Generalised processes at the margins of tectonic plates

along the west coast of the North American continent. The earthquake activity in California is related to the movement of these two plates. It is assumed that the plates have changed their position relative to each other and that effectively the constructive margin in the Atlantic has moved the American landmass away from the Eurasian and African continents. The movement of the half a dozen or so major plates and many smaller plates is not known exactly, and although there is acceptance of plate tectonic theory in principle there are many unknown details of precisely what is happening and why – the detailed processes in the system are largely unknown and unexplained.

The movement of the tectonic plates is only one subsystem within the geosphere, yet often the landscape impact of the processes in this subsystem is very considerable. Areas of present and former volcanic and earthquake activity may be related to the movements of these plates and the devastating local effects of many earthquakes result from large scale energy transfer at plate edges (see Chapter 15). Although the movement of the plates is gradual and continuous, often the energy release is very rapid so that, as found in other processes of landscape formation, the gradual and continuous change is suddenly supplemented by a short period of very rapid change.

The tectonic forces acting on the lithosphere affect the structure of the land surface, but activities within other subsystems of the geosphere also act on the rocks of the crust to produce the variety of land forms. Weathering processes cause the rocks to break down *in situ* and the loosened material may be eroded, transported and redeposited according to a variety of factors, resulting in the evolution of land forms through various stages (see Chapter 13). For the moment we shall consider briefly only how rocks are broken down. Two major types of weathering may be distinguished. *Mechanical weathering* processes involve the physical disintegration of rock, whereas in *chemical weathering* decomposition is more usual. Often processes of both types are active together, but usually chemical processes are dominant. The major division of processes in mechanical weathering is into pressure release or dilation, thermal expansion, crystal growth and biotic activity. When rocks which have been formed underground become exposed at the surface, then pressures are released within the rocks due to the removal of the weight of the overlying rock. With dilation the exposed rock may crack and effectively expand at the surface. The joints and cracks become weak places in the rock and other forms of weathering take advantage of the points of weakness. Weathering by thermal expansion, although with a long history of study, is probably a very minor form of mechanical weathering. The assumption is that, in areas such as tropical deserts, rock exposed to a wide daily temperature range expands and contracts and pressures are created which induce disintegrating stresses; however, laboratory testing of this hypothesis generally has failed to prove it. In fact the third process, that of crystal growth within a rock, is a much more important form of weathering. The most common example is when water freezes and the change to ice involves a volume change of just over nine per cent. Well-jointed rocks (with joints perhaps resulting from pressure release), for example chalk and limestone, are most affected by ice shattering, and the disintegration process can be quite rapid.

Finally *biotic weathering* means the capacity of plant roots to prise open bedrock joints, so increasing disintegration.

Chemical weathering processes are much more complex, with a great variety of chemical reactions occurring both underground and at the surface. When dissolved oxygen is in contact with minerals in the rock, oxidation may occur with the combining of the oxygen with elements such as iron, magnesium and calcium. Weak carbonic acid may be formed with carbon dioxide in solution and this weak acid is capable of dissolving some minerals. Water, on its own, can also be responsible for chemical weathering through the process of hydrolysis. With certain rock types, such as limestone, these chemical weathering processes have a significant landscape effect. The main processes and effects are summarised in Table 3.2

The relative importance of chemical and mechanical weathering depends not only on rock type but also on climate. The general relationship to climate is shown in Fig. 3.4. Chemical weathering is most important in warm wet climates, where the processes are

Table 3.2 Weathering processes: a summary of the main types

A Physical or mechanical weathering

Process	Effect
Sheeting, unloading, spalling	Release of pressure by removal of overlying rocks. Spalling forms irregular shapes.
Crystal growth	Includes salt and frost weathering: expansion of crystals exerts pressure. Also volume changes during chemical alteration.
Insolation	Temperature changes at surface: cleavage, cracking.
Fire	Forest fires leading to cracking as above.
Moisture swelling	Changes in volume.
Wetting and drying	For instance, at water level on shore platforms.
Cavitation	Bubbles in turbulent water leading to collapse.
Abrasion	Friction of boulders or impact.
Mechanical collapse	Following undercutting.
Colloid plucking	Clay film dries out, plucking out grains.
Soil ripening	Evolution of fresh alluvium: soil colloids lose water and volume, becoming sticky and then solid.

B Chemical weathering

Process	Effect
Solution	The first stage of reaction of water with rocks: amount of change determined by solubility of material and amount of water passing. Solutions may also lead to precipitation when saturated. Solute concentration in stream reflects rocks in basin and rates of percolation.
Oxidation/reduction	Adding or removing oxygen. Common form of natural weathering via oxygen dissolved in water. Oxidation mostly in aerated zone above water table; also by bacterial action. Reduction in waterlogged sites, where red/yellow oxides change to green/grey forms.
Carbonation	Reaction of carbonate, bicarbonate ions with minerals: often a step in the weathering process, including breakdown of feldspars. Abundance of CO_2 in soils aids process.
Hydration	Addition of water to mineral: important in forming clay minerals; often with large volume change. Prepares mineral surfaces for carbonation/oxidation.
Chelation	Chelating agents extract metal ions from organic chemical structure: assist leaching from humus.
Hydrolysis	Chemical reaction between mineral and water, related to hydrogen ion concentration of water: increasing concentration makes silica dissolve.

C Biotic weathering

Process	Effect
Particles broken	Animal burrowing; growing roots exerting pressure.
Transfer, mixing	Animal movement: materials moved to areas of different process.
Simple chemical effects	Solution enhanced by respired carbon dioxide.
Complex chemical effects	For instance, chelation.
Soil moisture effects	Roots, humus holding moisture; shading by plants.

Ground temperature	Shade; fermentation increasing heat and activity rate.
pH effect	Respiration and absorption by plants affecting pH.
Erosional protection	Less exposure, lowered weathering rates.

Source: Table 5.1 in M. J. Bradshaw, A. J. Abbott and A. P. Gelsthorpe, *The Earth's Changing Surface*, Hodder and Stoughton.

usually more rapid. Frost action is most important in cold moist climates. The combinations of these relationships result in different types of weathering regimes in various climatic areas. The weathering subsystem of the geosphere consists of a mix of processes which vary with rock type and climate. As such it is typical of many other subsystems in geomorphology concerned with the erosion, transport and deposition of material.

The biosphere

The subsystem contained in the fourth major earth system, the biosphere, are those which relate directly to the activity of plants and animals, including humans. Where atmosphere, geosphere and hydrosphere provide energy, water and nutrients in the correct balance, then life can be sustained. Particular points of balance give rise to different biotic communities, and the complex of plants and animals in a specific location is termed an *ecosystem*. Ecosystems have inputs of matter and energy, but also provide these within the system, using some and losing some through the system output. Within many ecosystems *photosynthesis* is a major energy and material forming process, where water plus carbon dioxide plus light energy give carbohydrate and oxygen. Photosynthesis may be viewed simply as a method of storing solar energy. This energy input is then used in growth, and as the energy is used up *respiration* occurs. This is the reverse of photosynthesis, and carbohydrate and oxygen react to produce water, carbon dioxide and chemical energy. Photosynthesis and respiration therefore are linked in a very simple cycle system (see Chapter 12) in which the input is solar energy and the output is chemical energy, and the result of which is growth. This simple statement is an oversimplification of some very complex biochemical processes, but nevertheless

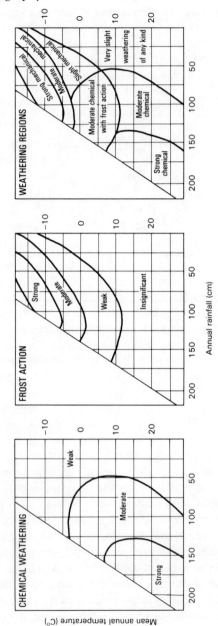

Fig. 3.4 Differences in intensity of weathering processes in different climates

points to the importance of cycles in the ecosystem. Photosynthesis is the basic process within the ecosystem and production in this primary process is sometimes measured to compare the capacity of different ecosystems to sustain growth. The net production by photosynthesis is the *biomass* measured as the dry weight of the organic matter produced. This can be stated for a single plant or animal, but it is more useful to measure it for a fixed area, for example a square metre, over a single year in an ecosystem. Forests have the largest biomass measured in this way, with equatorial rainforest producing about four times more than grassland. Table 3.3 shows some comparative values for typical ecosystems. The two most productive environments are the equatorial rainforest and tidal estuaries. The greatest range of values is associated with agricultural land, and this reflects the considerable differences in agricultural methods and subsystems of agricultural production. Plant biomass production from extensive grazing, for example, is much less than from market gardening, but this does not necessarily mean that economic profits are larger from the more ecologically productive subsystem.

Table 3.3 Biomass in grams per square metre per year in several types of ecosystem

Generalised ecosystem	Average biomass	Range of values
Equatorial rainforest	2000	1000–5000
Freshwater swamps and marshes	2000	800–4000
Mid-latitude forest	1300	600–2500
Mid-latitude grassland	500	150–1500
Agricultural land	650	100–4000
Lakes and streams	500	100–1500
Extreme desert	3	0–10
Tidal estuaries	2000	500–4000
Continental shelf	350	200–600
Open ocean	125	1–400

Each ecosystem with its characteristic biomass also has its own particular combination of species. Ecosystems change through time as organisms within the system react with and alter the physical environment. *Succession* in a ecosystem is the sequence of changes that occurs in a plant and animal community as it develops at a particular site. The initial *prisere* stage, such as breaking down bare rock surfaces, may take a long time under natural conditions, but may be speeded up by man's intervention. As plants become established, the *seral* stage is reached when a complex ecosystem is achieved but it has not yet reached the final *climax* stage. At the ecosystem climax the processes of change are only those of regeneration, not of species and community development, and a state of dynamic equilibrium is achieved. The climax is a final steady state. The times from prisere to climax stages vary very considerably. Studies on new sand dunes suggest a period in some cases as short as 100 years, whilst other studies on lava flows in temperate climates suggest that it may take over 1000 years to reach a climax vegetation. The change of dominant species is considerable over such a period, from – for example – beach grass, through herbs, coastal scrub to full forest cover. The biomass increase during the succession is large and species variety also increases. In other cases, however, where some particular aspect of atmosphere, hydrosphere or geosphere is a serious constraint on ecosystem development, then the time from pioneer to climax stages may be relatively short and the ecosystem may change little. If the temperature constraint is particularly strong, for example, the pioneer plants are unable to modify the ecosystem environment and climax and prisere plant communities are similar.

The three stages represent an idealised pattern of succession in an ecosystem, but often this sequence is disturbed, for example, by fire or flood or man. When sequential development starts again it is termed a *secondary succession* and usually the rate of change is more rapid than in a primary succession. In the eastern USA, for example, after the abandonment of farm lands it takes only about 200 years to regrow the climax forest community. The climax stage is the most complex ecosystem that the physical environment at the site can support, and makes the most efficient use of solar energy and available mineral nutrients. The species mix in a climax community, the time required to reach the climax stage, and the degree of stability at this stage, all depend on factors of the atmosphere, hydrosphere and geosphere such as climate, soil moisture, rock type and slope of the land. On a world scale there are broad variations in vegetation climax ecosystems paralleling the broad world climatic belts, but on a smaller scale there are also variations over quite short distances. The difference in ground water between valley side and valley floor may be sufficient, for example, for climax vegetation to differ from woodland to marsh and for quite different ecosystems to be supported on this vegetation. Similarly north and south facing slopes may be sufficiently different in their microclimate to support different ecosystems. In many parts of the world the most significant environmental factor determining the form of the climax community is the activity of man. Often man's effect is to halt the succession before the climax community develops. For example, grazing domestic animals prevent tree regeneration and provided this arresting factor is continued then a *plagio-climax* or *disclimax* grassland community will result.

Although the vegetation-based ecosystems depending on the photosynthesis of solar energy are complex and varied, the human systems within the biosphere are even more complex and varied and depend on a wide range of energy sources. The system of land-use generation in the city, for example, gives rise to a complex pattern of land use (see also Chapters 10 and 13). In most cities in Europe, the intensity of urban uses grades out from a high point at the city centre to a zone of low intensity around the rural–urban fringe. This oversimplification serves as a first step toward understanding city systems. It is then possible to build on this simple idea and elaborate it until it is a more accurate representation of what we know cities to be like. We can measure land-use intensity in several ways, but a fairly straightforward way is to consider economic rent levels. Fig. 3.5 shows the pattern of economic rent levels out from the city centre. Because the city centre is accessible to people from the whole city it is a much sought after location for the types of economic activity which involve direct interaction amongst people from all over the city. Large banks, large shops and major offices all try to locate in the city centre because of its accessibility to the city population. Consequently rents are high, and only activities capable of paying high rents locate in the central business district. Within the built-up area of the city there are other highly accessible points, perhaps where suburban rail and/or other public transport facilities have a major junction. These locations are particularly attractive to activities which do not have a city-wide appeal but which serve a major sector of the city. Here we might expect, for example, to find smaller branches of the big central-city department stores. Scattered through the city are places of slightly lower accessibility and

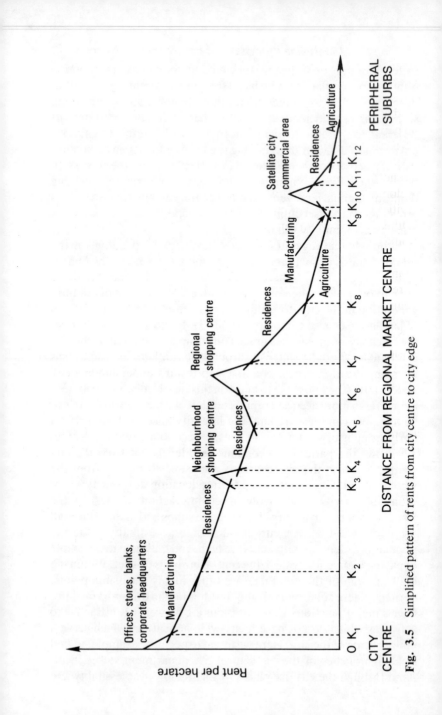

Fig. 3.5 Simplified pattern of rents from city centre to city edge

consequently lower rent, where local shops are found. If we consider the variation in rent through the city then there is a peak at the
centre, a few lower peaks and several smaller peaks, all superimposed on the gradual general decline outwards from the city centre.
The pattern of rents associated with accessibility levels in the urban
area is one factor underlying the system of land uses in the city. The
activities of land-use planners attempting to manage the system is
yet another factor affecting, for example, the pattern of office
development in the central city, and may effectively force some
activities to nodes elsewhere in the city.

Other examples of significant planning decisions are the imposition of a Green Belt around the city to control urban spread (see
Chapter 17), or the improvement of the urban road network, which
enhances the accessibility of certain nodes at the expense of others.
There are other factors, other inputs, which have to be included to
obtain a full understanding of the urban land-use system of which
the eventual land-use pattern is the output. The processes by which
the various factors, such as urban rent level and planning authority
decisions, generate the land-use pattern are the internal processes
of the system.

Four basic environmental systems can be distinguished: the
atmosphere, hydrosphere, geosphere and biosphere. Each contains
many subsystems, but it is important to remember that the four
systems interact with each other. The various agricultural subsystems, for example, interact with all four major systems. Farming is
influenced by physical conditions such as climate, soil water content
or slope, and by human factors such as the cultural and economic
attributes of the farmer.

An alternative classification of environmental systems

In order to show these interrelationships geographers sometimes
impose another artificial division on the systems comprising the
landscape. In a beach system, measures of form such as slope and
grain size are associated with one another. In American cities
dissatisfaction with associated social conditions arising from unemployment, inadequate housing, poor schools, high crime rate and
poor police protection leads to the formation of ghettos. The beach
and the ghetto are examples of *morphological systems* describing

associations of forms within the landscape. *Cascading systems* are defined by the energy flow within them. The outputs of one system become the inputs for a second and so on in a cascading chain. The thunderstorm subsystem described earlier can be viewed in this way and many subsystems in the economy are of this type. Outputs from agriculture form inputs for the industrial subsystem, which in turn produce outputs. These become the inputs for wholesalers, whose outputs are the inputs for the retail subsystem. *Process-response systems* relate the process-dominated cascade system with the description of form or morphology in the morphological system. The form of the landscape is considered to be a response to the processes operating in the landscape. For example, a morphological system of weathering in which variables such as climate, rock type, slope and so on, are associated together can be combined with a cascade system of the processes of debris movement downslope to create process-response system of the processes and forms of hill slopes (see Chapter 10). Process-response systems are more complex than either of the first two types. Finally, there are *control systems* in which key processes in a process-response system are controlled and managed. The urban land-use system is at first sight a process-response system, but land-use decisions are often controlled by a land-use planning authority and so in reality the system is a control system. Agricultural systems also are of this type and often have more than one process subject to man's control and management.

The main drawbacks to the system approach are the difficulty of defining boundaries and of working within such artificial divisions of landscape. The system approach is useful in so far as it creates a framework for the understanding of the processes of landscape change. Furthermore this approach brings together the study of physical and human geography which otherwise are often considered as distinct subjects.

In this and the previous two chapters several ways of looking at the landscape have been introduced. The landscape not only changes from place to place but it is also in a state of perpetual change, both through gradual evolutionary and through rapid revolutionary processes. To describe the landscape we can use ideas of regions, networks and nodes, but geographers seek to explain why the landscape looks like it does as well as to describe what it looks like. To explain the landscape it is necessary to study the processes operating, including the activities of man as he uses the resources present in the landscape. One approach to this study of processes is through the idea of systems, with its cycles and successions of landscape change and the interaction of processes with each other.

PART TWO

Tools of the Geographer

4

Maps

The basis of the study of geography is the description and explanation of the landscape, and it is possible to use different types of language to describe landscape patterns and processes. Most commonly we use literary languages to describe what we see, and a wide range of terms – such as valley, central business district, corrie, rent gradient – have been coined to help and to sharpen our descriptive powers. Alternatively we can use mathematical language to describe landscape; for example, a system of formal equations may be used to describe the journey patterns in a city, or the erosive power of a glacier. The third type of 'language' used by geographers is that of maps and diagrams.

Specific and relative locations

In the main, maps are used to describe the specific location of a feature within a standard framework of reference. The commonly used framework of reference is that of latitude and longitude, usually measured in degrees and minutes, but with even more accuracy if desired. The position of San Francisco, for example, can be defined as 37 40N, 122 25W or Sydney as 33 55S, 151 12E. Other frameworks of reference may be used, and many countries have a basic grid superimposed on them. The National Grid in Britain, for example, allows precise definition of location. Thus the small town of Lampeter in West Wales may be defined as at a position of 52 7N, 4 5W on the latitude and longitude grid, and at SN 578 481 on the National Grid of the Ordnance Survey. It must be remembered that all such grids are arbitrary and man-made and that each has a defined origin. The location of a feature is defined in respect of its relationship with the origin of the grid. For latitude and longitude this origin is the intersection of the Greenwich meridian with the equator, and in the case of the National Grid it is at a point west and south of Land's End. Fig. 4.1 shows the latitude and longitude grid and the National Grid for the United Kingdom. The 400 East National Grid line corresponds with the 2° West line of longitude.

Maps also show the relative location of features in respect of each other. Fig. 1.3 of tinplate works in South Wales shows, for the earlier dates, a clustering of features. Location in respect of a national or international grid is relatively unimportant to the information shown on this set of four maps. The important features of the distribution described by the maps are, first, the reduction in the number of marked locations and, secondly, their clustered pattern. The maps show the relative locations of tinplate works with respect to each other rather than their latitude and longitude or National Grid location. A variety of cartographic methods are available to display particular sorts of locational relationship. It is possible to describe gradual spatial change in a landscape, as for example showing changes in height with contour lines, showing regional distributions, or individual locations.

Topographic maps

It is useful to differentiate between general topographic maps, which provide a summary description of the total landscape, and specific thematic maps, which describe a particular set of features within the landscape. The topographic map series are the work of specialist surveyors, whilst thematic maps are more usually the work of geographers and cartographers. Most countries have general topographic map series at different scales, with some sets probably covering the whole country. Two common scales for such general series are 1:250 000, and a more detailed series at 1:50 000. When scales are stated as a *representative fraction* we know by exactly how much the map size is a reduction of actual size. More detailed maps are sometimes available, even to a scale of 1:1000, but the areas covered by such maps are small. The Ordnance Survey in the United Kingdom provides a map series more comprehensive than is available for any other country, both in range of maps and detailed coverage. Geographers in Britain sometimes forget the relative lack of map coverage elsewhere! For large parts of the

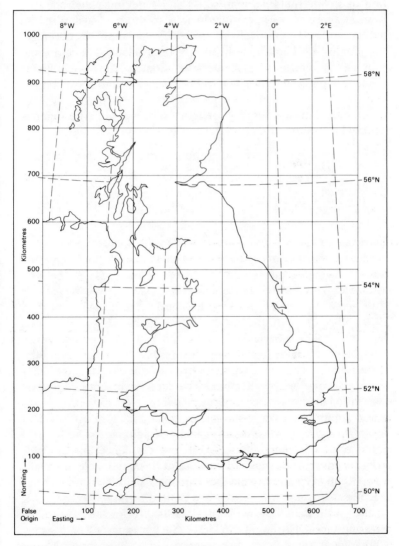

Fig. 4.1 The latitude and longitude grid and National Grid of the Ordnance Survey for Great Britain

world there are no maps of more detailed scale than 1:2.5 million, and even in some parts of Western Europe maps of 1:50 000 are not available. There have been several attempts to organise the production of an international map series at 1:1 million scale following general agreement in principle in 1891 to produce such a set. In excess of 2000 maps will be necessary to complete coverage. Perhaps by 1991 half the sheets will be available – such are the problems of international cooperation. Some of the developments in satellite photography which will be discussed in the next chapter may in the future call into question the need for this type of international map series. Two general objectives underlie an international map series:

1   The provision of a general purpose map to allow a comprehensive study of the world for pre-investment survey and economic planning.
2   The provision of a base map which can be used in the preparation of thematic maps.

The same broad aims underlie national topographic map series.

The Ordnance Survey map series in Britain dates from 1801, when a 1 inch to 1 mile (1:62 360) map of the county of Kent became available. This was the precursor to successive map series at this scale – the 'New Series' in 1840, the Third Series in 1893, Fourth in 1919 and so on, until the Seventh Series, introduced in 1959, was replaced by a First Series of 1:50 000 in the 1970s. A Second Series 1:50 000 is becoming available gradually sheet by sheet during the 1980s. Each series has a characteristic style and reflects the map-making technology of the time. The successive series provide a fascinating picture of the changing landscape over 150 years. There is also a national coverage in the larger scale maps at 1:25 000 and 1:10 560. The 1:10 560 scale is gradually being replaced by 1:10 000 and a process of continuous revision by Ordnance Survey field-workers ensures that up-to-date maps at this large scale are available. At an even larger scale there are maps of 1:2500 and plans at 1:1250 scale, though the latter are limited to larger towns. At a smaller scale than 1:50 000 is the 1:250 000 series, and two maps covering Great Britain at 1:625 000. Such a range of map scales allows geographers to choose and use a scale appropriate to the information they wish to display.

The range of landscape features shown on the maps of different scales varies considerably. On the Ordnance Survey maps, for example, individual buildings are shown at 1:10000 but not at smaller scales. The 1:25000 is the smallest scale to show field boundaries, whilst the 1:250000 series inevitably generalises many landscape features and shows few administrative boundaries. The decision on the choice of landscape features shown on a map also varies from country to country and a comparison of two or three 1:50000 map series for different countries quickly shows the variety of map content amongst national series. Although produced to a specified scale, many of the specific features on a topographic map are shown schematically and not to scale. On maps on a scale of 1:25000 and smaller, roads and rivers are seldom shown to scale. The accurate portrayal of size of such features at these smaller scales would make them so small as to be barely visible. With larger scale maps other problems occur in actually deciding whether a trickle of water is a stream and whether it should be shown as such. At a smaller scale, say 1:50000, only rivers and streams above a certain size can be shown and map makers have to decide what to include and what to leave out.

Although topographic maps provide a descriptive language of the landscape, it is a subjective language and a generalised picture that is produced. In order to communicate this picture specific techniques are used, and as with ordinary written language where words change through time, the ways of portraying landscape on topographic maps change through time. A variety of techniques is available for showing landscape height and shape, or relief and the various relief features such as spurs, cols or concave slopes. The simplest technique is to plot *spot-heights* on a map. A dot and associated value indicates the height of the point above mean sea level. A second method is by *contour lines*, which join points of the same height. The height interval between adjacent contour lines is a critical factor in the relative success or failure of this method in describing landscape relief. Thirdly there is *layer shading*, whereby all land between two particular heights is shaded in the same way or in the same colour. The technique may be misleading in suggesting that all the land of a single shade is a constant height, rather than ranging between two limits. Fourthly there are *hachures*, which are lines drawn down a slope in the direction of the steepest gradient, usually drawn closer together where the slope is steeper. Fifthly, a three-dimensional effect may be created by the addition of *hill-shading*, which tries to show the lighted and shaded parts of a landscape assuming an oblique light at the top left corner (usually the North West but not always) of the map. Other less commonly used methods are also available, together with the use of specific symbols for particular relief features; for example, cliffs can be shown by wedge-shaped black lines. Most topographic maps use some combination of these methods for showing the relief of an area. The 1:50 000 Ordnance Survey, for example, uses a mixture of spot-heights, contours and specific symbols. It is interesting to compare this range of techniques with those used on the Ordnance Survey 1:250 000 series.

Thematic maps

The same range of techniques may equally well be used in the production of thematic maps, for example to show the relationship between the landforms in a region. The distinguishing feature of thematic maps, however, is their concentration on one aspect of landscape. Some thematic maps are produced by national carto-graphic agencies and comprise map series at different scales. Geo-logical maps are of this type, and some countries also have a full cover of soil, vegetation and climatic maps. In some instances thematic maps have a particular set of symbols and colours to describe the feature being analysed. In synoptic weather maps some conventions for showing the frontal patterns, wind and temperature characteristics, are recognised and used internationally.

When the landscape is defined as patterns of points, networks and regions, then there are individual mapping techniques to deal with each type of building block. Point patterns are usually mapped by precisely located symbols. In its simplest form this might be a simple dot map, as in Fig. 1.1 showing deserted villages in Eastern Eng-land. One dot represents one village on a map with a scale suitable to show the area containing these places. An extension of this type of map involves the use of symbols of varying size, with the symbol (often a circle) proportional to the size of the feature located at the point. Thus in Fig. 1.1, if we had data on the maximum size of these villages, then it would be possible to plot a series of circles proportional to population size to show where total population loss was greatest and where it was least. More complex data can be shown by dividing the symbol in proportion to the ingredient elements. In a map of the number of new houses built in each town of a region, a proportional circle map could be prepared with circles proportional in area to the total number of new houses. It would be possible to show for each town the share of new houses built by local councils and the share built by private builders by dividing or sectioning the circle in proportion to the shares. Even more complex symbols may be plotted at particular points, as for example in complex diagrams showing the age and sex structure of a population. Individual diagrams, or population pyramids, showing the proportion of males and females in each age class could be calculated for each settlement in a region and the appropriate diagram drawn at the correct location on the map. Whether the symbol is a simple dot or a complex diagram the basis of this type of map is the same each time: the information about a point in space is communicated by a symbol on the map.

The mapping of information on networks can be carried out by a simple line-drawing of the network; or if the information is more complex a variety of other techniques can be used. In addition to showing the extent of connectivity, the map may also show the quantitative value of the flows within a network. This is usually achieved by making the width of the line proportional to the quantitative measurement of flow between selected points on the network. For example, this method can be used to show the quantity of goods transported within a railway network, with the network shown in detail and the width or thickness of lines being proportional to the volume of goods carried. International trade flows can be shown by a system of arrows, where the width of the body of the arrow is proportional to trade volume or value. Sometimes a single line represents a fixed quantitative flow value, with the actual value shown by using several parallel lines. Thus if a single line represents a flow of 1000 people, then a flow of 4800 people is represented by five parallel lines.

Mapping of regional information is carried out in several ways, but two main techniques may be isolated depending on whether the regional division is into formal regions or functional regions. The use of *choropleth* maps is common in illustrating information concerned with formal regions. In this approach a shading scheme is used to show regional values. The range of different values over the region is divided into several categories, often five or six regions are assigned to a category, and shadings are devised with a particular shade for each category. Several problems occur in the construction of these maps, particularly in respect of the selection of categories into which the values are divided and also in the choice of shading system. It is often possible to present quite a different visual impression by using different class intervals to divide the data into categories. There are no hard and fast rules on ways of dividing up the data. Some geographers prefer to divide the values into equal groups, whilst others prefer to study the actual values and try to devise a scheme which shows groups in the data. Table 4.1 shows three possible ways of dividing up a hypothetical series of twenty regional values. The values have been ranked for convenience.

Table 4.1   Alternative divisions of a series of values to be mapped using a choropleth technique

Sequence of ranked values	Four equal divisions (quartiles)	Division at 100, 75, 50 and 25	Isolation of extreme values and 'natural' breaks
170	1	1	1
94	1	2	2
90	1	2	2
88	1	2	2
80	1	2	2
73	2	3	2
72	2	3	2
53	2	3	3
53	2	3	3
52	2	3	3
48	3	4	3
48	3	4	3
48	3	4	3
36	3	4	4
34	3	4	4
34	4	4	4
28	4	4	4
24	4	5	4
4	4	5	5
1	4	5	5

(N.B. Numbers refer to classes into which the values are placed.)

The second set of decisions to be made in preparing a choropleth map relates to the shading scheme. Choropleth maps may be used to show non-quantitative data such as land use or rock type, and then an ungraded series of shading patterns may be used, as on the geological maps. Alternatively, when quantitative data are being shown, a graded series of symbols may be used, usually with the densest shading pattern showing the highest values to be plotted. Spacing between lines in a shading system may decrease as the values become smaller, so that a visual impression of the quantitative data is provided. It is also possible to use different densities of dots, with higher densities being associated with higher values and vice versa.

The plotting of data associated with functional regions involves a quite different technique termed *isopleth* mapping. In this type of map quantities are indicated by lines of equal value. The contour map is a form of isopleth map. In all isopleth maps it is assumed that a continuous surface is being shown and that individual contours on the surface are plotted. On a map showing the proportion of people visiting a town to shop, the trading region of the town may be mapped by isopleths showing percentages of shoppers from different areas. The isopleths would form a series of very generalised rings around the town, with the distance between the isopleths showing the decline of retail influence of the town centre with distance away from it. The main difficulty associated with the compilation of isopleth maps arises in *interpolating*, or calculating, the values between those which are exactly known. Particular values may be known from some interview shopping survey. Alternatively, with an isopleth map of temperatures the particular data from weather stations are known. From these particular values it is necessary to calculate intermediate values, which can then be joined up to produce isopleths. The simplest interpolation procedure would assume that the isopleth of 20°C would pass midway between stations recording 15°C and 25°C, but more sophisticated interpolation methods are possible.

The purpose of all these techniques is to show the spatial distribution of some thematic feature of the landscape. The presence of settlement might be shown by a dot map; the location and flows of a railway system by a flowline map; population density by a choropleth map; and rainfall by an isopleth map. The different techniques are used for communicating different types of information.

Recent developments in map making

The mapping techniques considered so far in this chapter have a long history of development and design. Two relatively recent innovations are topological maps and computer drawn maps. In topological maps, more correctly termed *cartograms*, areas are shown as proportional to the size of the variable being mapped. This technique provides a very vivid way of showing data with a wide range of values, for example population or national income. The aim is to distort the map as little as possible to preserve recognition, but at the same time to produce sufficient distortion to stress the differences in the data. In a population cartogram, countries of small area but large population may be emphasised, compared with countries of large area and a relatively small population.

Other recent developments in both topographic and thematic map making have involved the use of computers to store and process information. The map drawing is then carried out by computer-controlled pens and printing machinery. The mass of information present on a topographic map can be stored in modern computers, and as landscape changes occur and maps require updating, so it is both fast and easy to update the file of computer-held information. The calculations necessary before a thematic map can be produced are easily produced by a computer program and several types of drawing device can be connected to the computer to automate the map production process. Symbol, flowline, choropleth and isopleth maps can all be produced in this way. The speed of map production has the added advantage that different scales and symbols may be tried out on the one set of data to determine the best technique of communicating the information.

As well as the large computers and sophisticated equipment used for such complex programs, there are also micro-computers such as the PET, SPECTRUM and APPLE, which can be used to display maps and information. The general availability of these small computers allows many types of cartographic experimentation. For example, *dynamic* maps can be used to show landscape change using similar principles to cartoon production. Maps of the outline and form of a river-mouth spit can be drawn and then redrawn using data collected perhaps at weekly intervals. The map sequence can then be displayed in a sort of time-lapse cartoon, and the changing shape of the feature clearly illustrated. There are many other features of landscape change which can be shown in this way, from long term changes in urban population density to the changing pattern of river meanders across a flood plain. This aspect of map making is currently in its infancy, but the techniques will be increasingly used to describe the changes in the landscape.

5

Air Photographs and Remote Sensing

Maps have important limitations. Topographic maps, such as the Ordnance Survey 1:50000, depict a carefully chosen mixture of landscape features. The choice of features for inclusion varies from country to country, but it is impossible for any map to depict the total landscape. Moreover, since landscape is changing continually, any mapped depiction of contemporary landscape is out of date as soon as it becomes available. To overcome these difficulties aerial photographs are increasingly used by geographers, both as sources of basic information about the landscape and as analytical tools for the study of landscape change.

Origins and types of air photography

In the early nineteenth century photographs were taken from balloons, but they were generally for either military use or simply general interest. This captive balloon photography was used extensively during the American Civil War to make photo maps of troop positions. During World War 1 in Europe aerial photography was again used for intelligence, and considerable advances were made in the production, technology and interpretation of air photographs. Geographers began experimenting with air photograph analysis during the 1920s and 1930s, but the utility of air photography for geography became apparent in the post World War 2 era, when the advances in interpretation techniques made by the military became available for civilian purposes. During the last twenty years further advances have been made using earth-orbiting satellites, better cameras, improved interpretation techniques and alternative types of 'image' collection.

The products of air photography and related remote sensing methods are used by geographers in three different ways. First, the photographs or other types of image may be interpreted directly. Secondly, they may be used as a first step in the map production process. Thirdly, quantitative analysis may be carried out from measurements taken from the image. The three activities are not exclusive. When photographs are used in map production detailed measurements have to be made and images have to be interpreted before the photograph content can be summarised in a map. The analysis and interpretation of air photographs and other types of image such as those produced by thermal and radar scanners is a skilled and complex activity. It has developed into a specialism within geography, but as with map making it is usual for all geographers to have the basic ability to interpret them.

Materials, for all effective purposes, emit or reflect electro-magnetic energy, which is radiant energy moving at the velocity of light in harmonic wave patterns. The wavelength of this energy can vary, and the human eye and brain are capable of receiving and processing some wavelengths termed visible light. There are many other wavelengths invisible to the eye, and these can be processed by other types of receiver, producing an image capable of inter-pretation. Radar images are generated from energy of relatively long wavelength and images may be collected using radar scanners. Something more like the traditional camera can act as a sensor of wavelengths close to and including the visible wavelengths. Gener-ally, when solar energy is received at the earth's surface it is either reflected, transmitted, absorbed, emitted or scattered, depending on what sort of material it falls on. So different materials reflect or emit a different mix of energy wavelengths. By analysing this mix, sometimes called the *signature*, it is possible to identify the material from which the energy has come. In the visible wavelength the signature may be a colour, or in black and white photography a particular shade or tone. These signatures also exist for wavelengths other than visible light. The ability to emit infra-red radiation, for example, is linked to surface temperature. Cooler areas, often associated with higher water content, are more easily distinguish-able from warmer, often drier, areas using infra-red images than by using visible light images. A relatively recent development has been multiband scanners which are capable of receiving and processing several wavelengths and of keeping the images separate. The equipment often consists of a complete camera-type receiver with four or more individual and separate receivers each taking an image from the same piece of landscape, but each responsive to its own particular range of wavelengths. These multiband scanners have been used extensively in earth-orbiting satellites.

The traditional air photographs, using visible light, are still widely used by geographers and are generally of two types (see Fig. 5.1). *Verticals* are those in which the camera is pointed vertically downward, producing a photograph resembling a plan. *Obliques* are those in which the camera is tilted, so that the photograph looks as if it has been taken from a high building. If the camera is tilted so that the horizon is shown, then the photograph is termed a high oblique. The air photographs used for publicity and newspapers are often low obliques with no horizon. Interpretation is frequently much easier with obliques than with verticals, but more precise measurement and analysis is possible on vertical photographs, which have less variation in scale than oblique photographs. Most visible-light air photography now used by geographers is in the form of a sequence of vertical photographs with a considerable overlap (usually 60% in one direction and 25% in the other). This allows stereoscopic viewing, and also allows photogrametric techniques to be used in photo-interpretation and analysis.

Air photograph interpretation

The scale of the photograph is important in air photograph interpretation, and this depends on the height of the camera. Even on verticals the scale is not constant over the whole photograph, due to the increase in image distortion towards the edge. The approximate scale is often provided along the edge of the photograph along with the date, time, flying height, focal length of the camera and other information – all of which is useful in the interpretation process.

Some photograph interpretation and analysis can be carried out by sophisticated technologies involving automated pattern recognition procedures, but much interpretation by geographers is carried out visually by observation of an object's shape and size, its

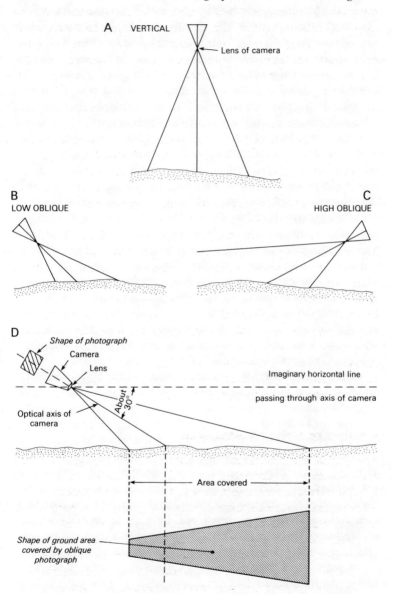

Fig. 5.1 Camera positions for vertical and oblique air photography

colour, the features with which it is associated, texture, pattern and often by the shadow cast by the feature. The shape of an object on a vertical photograph is usually its plan form, and sometimes familiar objects appear somewhat strange when viewed in this way. It can be difficult to assess the use of a building from its plan form alone, but features associated with the building may help to identify it. If it is a manufacturing plant the shape of the building alone might provide indications of use, as continuous production processes often require relatively long, narrow buildings. The mix of equipment, stock-yards, transport links and so on associated with the feature will help considerably with its identification. It is more useful, as a rule, to consider land-use units rather than individual buildings or features. The combination of buildings and their immediate environment is often easier to interpret than a single feature or building in isolation. Schools as buildings may be difficult to locate on a photograph, but in association with a hard-surfaced playground or even playing fields they become easier to identify. Through the pattern of wear of the grass on the playing fields it can be possible to distinguish rugby pitches, with wear along the side lines, from soccer pitches with wear in the goal area. Much interpretation depends upon consider-ing small details, such as the shop awnings on a row of buildings – thus distinguishing between shops and houses. The size and shape of objects are all important in this exercise.

Texture and pattern are also important, particularly in agricul-tural and vegetational studies based on aerial photography. The texture and pattern of oak woodland seen from above will be quite different from that of spruce or larch. Regular patterns of trees might suggest an orchard, and proof might then be sought by looking at the associated features – buildings, pallet storage and so on. In the case of playing fields, the texture of a worn area of grass looks more patchy and lighter in appearance than an unworn area. The texture of the image is a response to the amount and evenness of the light reflected from the object. Larger amounts of light are reflected from smooth surfaces than from rough ones, and so on black and white photographs the smooth surface appears lighter. The smoothness and roughness of a surface has a noticeable effect on the photographic image, so it is possible to pick out easily a regular path across a field. The path will appear lighter because the bent grass reflects more light than undisturbed grass. Often on photographs of pastureland it is possible to see cattle tracks converging on a break in the boundary fence of the field. The study of the combination of texture and pattern is especially important in vegetation surveys where differences in the mixture of bracken, heather or cotton grass within rough pastureland result in tonal differences on the photograph. Variation in tree species and age produce identifiable differences in the pattern and texture of the photographic image.

Shadows can be extremely useful in the interpretation of individual features. A vertical photograph of a chimney or oil derrick can present a difficult interpretation problem unless its shadow is recognised. The shadow cast by a tree can solve the problem of species identification. By measuring a shadow the height of an object can be calculated. The height of an object is related to the length of the shadow by the tangent of the angle of the sun (by the simple geometry of a right-angled triangle). Published tables are available from which the angle of the sun can be calculated. Difficulties can arise with this method, however, when objects are located on a slope and cast shadows either up or downslope, or if the shadow is cast on snow or even when the object itself is not quite vertical.

When measurements are taken from air photographs it is all important to know the scale. The scale of a vertical photograph may be calculated as:

$$\frac{\text{Camera focal length}}{\text{Altitude of the camera above the ground}}.$$

For example, with a camera of focal length of 150 mm, flight altitude of 2000 m above mean sea level, and an average ground elevation of 500 m, the scale of the photograph would be, with all figures in metres,

$$\frac{0.15}{2000 - 500} = \frac{0.15}{1500} = \frac{1}{10\,000} \text{ or } 1{:}10\,000$$

This scale would only be precise at a ground altitude of 500 m. In mountainous areas where the variation in relief, or *relative relief*, is large, the photograph is subject to considerable variations in scale. With a relative relief of 200 m in the above example, the photo scale will vary between 1:9337 and 1:10 672. With an object measured on the photograph as 1 cm this would mean a difference between 93 m and 109 m in the estimated real size of the object. If areas were being calculated, then considerable errors could arise in simple analyses in heavily dissected landscapes. Accurate measurement from air photographs is a complicated exercise and is the field of expertise of specialist photogrametricians. Geographers, although concerned with measurement, also use air photographs for landscape and land use description.

Stages in interpretation

Descriptive interpretation is usually based on a systematic and deductive approach to studying the photograph. Interpretation proceeds from consideration of the general patterns to the identification of specific features. This step-by-step approach can be illustrated by the accompanying list of procedures used in the interpretation of urban land uses:

1 Outline the built-up and urban area on the photograph.
2 Mark the major land and water transport routes passing through the urban area.
3 Isolate and mark the major airports.
4 For the built-up area outline subareas to show water-bodies, natural vegetation, relief features, etc.
5 Divide the built-up area into subareas with different street patterns.
6 Outline the old and new parts of the city, picking out any historic buildings, castles, palaces, etc.
7 Mark on the principal transport lines in the urban area.
8 Mark the minor land and water lines passing through the urban area.
9 Circle the major stations, terminii and transport interchanges in the urban area.
10 Outline the primary commercial subareas in the town centre and in the suburbs.
11 Outline the major industrial subareas, industrial estates and utilities such as power stations.
12 Outline subareas of warehouses and open storage.
13 Mark recreation areas.

14 Mark the major cemeteries and areas of allotments.
15 Outline residential subareas with different dominant house types including size of plot.
16 Circle the principal administrative and government buildings.
17 Outline the major schools and education centres.
18 Mark the secondary commercial centres.
19 Mark the isolated industrial plants.
20 Circle major features which at this initial interpretation stage are not identifiable and will be returned to for more detailed study later.

After this initial stage of interpretation it is possible to consider patterns and relationships of urban land uses in the city and to draw conclusions about the geography of the area being studied. At the second stage a particular urban activity and land use may be studied in more detail, using the same step-by-step procedure. In the study of industrial land uses, for example, it is first necessary to decide whether the activity is an *extractive*, *processing* or *fabricating* industry. *Extractive* industries are characterised by features such as excavations, mine winding gear, ponds, piles of waste, handling equipment such as conveyors, and heavy duty machinery; buildings are few and relatively small. *Processing* industries are divided into the three subclasses of mechanical, chemical and heat processing. Mechanical processing industries are those which change the physical form and appearance of raw materials, such as saw mills, ore concentration plants or grain mills. Characteristic features of these industries are facilities for the storage of large quantities of bulk materials and energy generating or transforming equipment such as boiler houses or electricity transformer stations. It is often possible to identify the precise industry from stacks of raw materials. Chemical processing industries are typified by many tanks or gasholders, often in a regular pattern; a considerable network of pipelines; and large, outdoor processing plant. Heat processing industries use heat to refine or reshape raw materials. Large quantities of fuel are needed and fuel dumps, waste piles, blast furnaces, kilns and complex buildings characterise this type of operation.

The third main group consists of *fabricating* industries, which use the output of processing plants to make or to assemble finished goods. Of the industrial activities this subgroup is the most difficult to distinguish in detail as much of the activity takes place inside general purpose factory buildings. Heavy fabricating industries, such as shipyards or bridge building establishments are relatively easy to identify because of the presence of heavy cranes and other outside equipment. Light fabrication industry is much more difficult to determine, and often only hints for specific identification are available from roof patterns, vehicles in the factory yard and detail of materials stored outside. Quite different industrial processes carried out within standard factory units on an industrial estate are frequently indistinguishable from one another from aerial photo analysis. Other types of survey and data collection methods have to be used to obtain specific information in these cases (see Chapter 6).

This step-by-step approach to interpreting air photographs, with individual photographic images being considered in terms of size, shape, colour, pattern, texture and associated features, is the basis of all interpretation efforts, and not only those relating to urban and industrial land uses. The study of land forms and physiographic features can be carried out in this way, with meander patterns in river flood plains or drumlin patterns in glaciated lowlands clearly visible on vertical air photographs. The step-by-step approach is also valuable in rural land use studies and the identification of crop types. In this connection an additional feature of air photograph interpretation becomes important: the date and time of day when the photograph was taken. A photograph taken in the spring will show a different landscape from one taken of the same area in autumn. Seasonal contrasts, which of course are absent from topographic maps, are particularly important in mid-latitude regions such as Europe and the USA, where marked and critical seasonal differences in climate exist. The date of the photograph is almost always provided in the information on the photograph margin. In intensively cultivated temperate areas, such as most of Britain, the seasonal change in vegetation and soil moisture is easily seen. In spring, field patterns appear sharp and distinct due to different stages in cultivation and crop growth. Mottled textures indicate differences in the soil moisture content. Even slight differences in the height of the ground can be identified from differences in drainage and soil moisture: higher, drier land appears lighter. In summer, photographs are dominated by the darker tones associated with maturing crops and taller ground vegetation. Soil moisture is often low, so bare earth appears light in colour and even in texture. In the autumn, field patterns again become distinct because the crops are at different stages of harvest. In winter, field patterns are indistinct and generally high soil moisture gives a more uniform dark tone to the ground. The low angle of the sun throws clearly visible shadows of vegetation and land forms.

Satellite images

With traditional air photographs it is unusual to have access to photographs of the same area taken at different times during the same year, but with satellite photography it is easy to obtain several photographs per year of the same locality. The two *Landsat* satellites, which were launched in 1972 and 1975, photographed the same area of the earth at exactly the same local time every eighteen days. All areas, except the polar regions, were covered once every eighteen days. The satellites contained a very high resolution TV camera and multi-spectral scanners capable of obtaining images from two wavelengths in the visible part of electromagnetic spectrum and two wavelengths in the near infra-red range. Each image taken covers an area the shape of a parallelogram with each side 185 km. Generally they are available at printed scales of 1:3½ million, 1:1 million, 1:500000 and 1:250000. The high resolution on these images is such that the character of an area 80 m × 80 m on the ground can be picked out on the satellite image, provided sensing conditions were optimal at the time of image collection. Even this high degree of resolution has been improved upon by sensors on more recent satellites which are capable of images resolved to 20 × 20 m ground areas. Thus it would be possible to pick out a moderate-sized school playground on one of these images taken from over 900 km away. With such high levels of resolution the amount of information transmitted by the satellite sensors is so enormous that information is created faster than it can be analysed and so sensors are only switched on for short periods. Otherwise the amount of information sent back to the receiving stations on earth would be totally indigestible, even by the largest computers.

The American space shuttle will help considerably in this respect, for it will allow a more selective programme of remote sensing of the earth, instead of the blanket coverage produced by the *Landsat* and other early satellites. With the space shuttle it will be possible to put into orbit specific equipment to look at a particular area, and then to replace the experiment or modify it on later trips. When the shuttle returns, users will be able to assess the data quickly, plan the next mission, and then return to analyse in detail the data from the earlier mission. Computer and other micro-electronic technology is vital to this satellite image assessment, interpretation and analysis. The geographer now has the equipment for detailed analysis and enlargement of selected parts of an image through a visual display unit. As with map interpretation, image and air photograph interpretation have moved into the micro-electronic age.

6

Surveys

Maps and aerial photographs are sources of data about the landscape from which geographers can select, analyse and interpret information in order to describe a variety of landscape features. To explain and understand these landscape features fully, however, it is often necessary to carry out survey work using scientific methods. Survey information may be mapped, mathematically and statistically analysed, and interpreted in words. Surveys are a basic form of data collection for geographers.

Types of survey

Surveys are used widely in all aspects of geography, but are particularly important in collecting information about people and their activities. They can be divided primarily into those in which questions are asked of a group of people, and those in which data is collected by visual survey. This division separates, for example, surveys in which a questionnaire is sent to people requesting information about their activities, and surveys in which people, or landscape features, are observed. Information about shopping activities can be collected by either technique; information on agricultural land use can be obtained either by asking the farmer or by visually surveying the fields.

Secondly, surveys are divided into census and sample surveys. In census surveys every person or feature is surveyed; in a sample survey representative people or features are surveyed and the data are assumed to be typical of the whole group. A survey might be undertaken of the time and length of the journeys to a particular school. In a census survey every pupil, teacher and other employee would be asked about their journey pattern, but in a sample survey only a proportion would be questioned. In any sample survey there will be some error, as the group of people chosen will not be exactly representative of the whole group. This error is called the *sample error*, and by taking different types of sample it is possible to minimise this. In physical geography, surveys using sample methods are common. In a study of pebble types and shapes on a beach, a sample of pebbles might be collected from near the waterline and another from high up on the beach. Then measurements of the pebbles would be taken and conclusions drawn about differences in size of beach material in the two areas. Obviously a sample would be taken in this case, but the way in which the sample was taken might well affect the results. A very carefully devised sampling scheme would have to be used to stop, for example, the collector of the sample simply picking the largest, or most easily seen pebbles. The sample could be *biased* if an incorrect *sample method* is used. Sampling methods are described later in this chapter.

Not all censuses and sample surveys used by geographers are carried out by them personally. Many of the major surveys are the responsibility of international agencies, such as the United Nations, or national government agencies such as the Office of Population Censuses and Surveys (OPCS) in the United Kingdom. Surveys of the number of people, their age and other vital statistics are typical of those carried out by national agencies, sometimes within a framework coordinated by the United Nations. The *United Nations Demographic Yearbook* presents the basic results of these data on an international basis, and from this single source analyses can be carried out of population numbers, density, sex, age, birth rate, death rate and many other measures of population.

Population censuses

In a coordinating programme the United Nations suggest that each country should carry out a census survey of its population on a regular basis. They have divided the type of information (shown in Table 6.1) that is collected into lists of *recommended* and *other useful* topics (the latter are shown by an asterisk in the table). These

Table 6.1 Topics suggested by United Nations for inclusion in population censuses

Surveyed Topics	*Calculated Topics*
Geographic characteristics	
Place of population at time of census	Total population numbers
Place of usual residence	Population by locality
Place of birth	Urban/rural division of population
Place of previous residence*	
Duration of present residence*	
Place of work*	
Personal and household characteristics	
Sex	Household composition
Age	Family composition*
Relationship to head of household	
Relationship to head of family*	
Marital status	
Age at marriage*	
Duration of marriage*	
Marriage order*	
Children born alive	
Children living	
Citizenship*	
Literacy	
School attendance	
Educational attainments	
Educational qualifications*	
National and/or ethnic group*	
Language*	
Religion*	
Economic characteristics	
Type of activity	
Occupation	
Industry	
Status (employer, employee etc.)	
Main source of livelihood*	

*Topics considered useful but not essential

topics, sometimes termed *dimensions*, the United Nations report suggests, 'have emerged after decades of census experience as of the greatest utility for both national and international purposes'.

Most, but by no means all, countries have a population census which involves a survey of the basic information listed in Table 6.1. Some countries, particularly those of Western Europe, Canada, the United States, Australia and New Zealand, have population censuses which cover many more topics.

In England, Wales and Scotland the first official population census was in 1801 and it has been repeated every ten years until 1981, except for a break in 1941. Additionally an extra sample survey was undertaken in 1966. With each successive census more elaborate questions have been asked and more information collected about the population. The 1801 census comprised simply a count of the number of males and females in each house and family, and the number of people working in agriculture, craft industry, manufacturing, trade and other occupations. Later censuses introduced new topics. Age of the population was collected in 1821, although it was optional for an individual to answer this question until 1851. In 1891 extra information was gathered on housing and the number of people occupying each room. The 1911 census extended the information required on marriage and fertility. Since 1921 the number of questions on the census has remained more or less the same, but changes in the way the questions are asked and the exploration of specific topics has generally meant that increasing amounts of geographically useful data have resulted from successive censuses. The censuses in the United Kingdom seek to provide data for a basic description of the number of people, their economic activities and how they live. As such it fits well into the United Nations' definition of a population census as, 'the total process of collecting, compiling, evaluating, analysing and publishing demographic, economic and social data pertaining, at a specific time, to all persons in a country or well-defined part of a country'. The social data referred to in this quotation include information not only on individual people, but also household and family groups and the housing that these groups use. The National Population Censuses have had their purpose extended from being simple counts of numbers of population, and now provide a general social survey.

Population census agencies are not the only organisations to carry out census surveys of society. In the United Kingdom the planned official 1976 census was cancelled and several local government land-use planning authorities undertook census surveys in the mid 1970s. These surveys are often accessible to geographers, and although they do not provide a national coverage, they do provide sources of recent information for several large British cities. Comparable agencies in other countries also carry out censuses and surveys of the social characteristics of the population. The results of the 1981 UK census will allow up-dating of the local surveys. The task of undertaking a census survey is very considerable. Great care is required in designing the survey documents and phrasing the questions. Because of the large scale of a social or population census which involves contacting every household in the study area, a variety of techniques of sample survey has been developed which reduces the number of people surveyed but which, of course, does not reduce the difficulties associated with the phrasing of questions. The use of sample survey methods introduces extra problems of choosing and contacting the specific individuals who comprise the sample to be interviewed.

Sampling methods

In geography, sample surveys are used to explore in depth some aspect of society, human behaviour, or landscape features. Such surveys can complement census of population material by collecting information on topics such as income and family expenditure, which are seldom covered in large scale censuses. Alternatively they can deal with topics well outside the usual scope of official population censuses and collect data, for example, on people's shopping behaviour (where and why they shop), on recreational activities or even on what sort of landscapes people like and dislike. Sampling is defined as the selection of part of a total in order to represent the total. In any sampling procedure the total has to be split into *sampling units*. These may be natural units, such as people or families in social surveys or specific landscape features in geomorphological surveys (see for example Fig. 10.3). Alternatively they may be artificial units, such as areas on maps. The units in both cases must be capable of exact definition. The total number of sampling units is called the *population of sampling units* and from this a *sample* is taken or *drawn*.

The sample drawn must portray the total as accurately as possible, and two features are important in this connection: the way the sample is drawn, *sampling method*, and the size of the sample, *sample size*. There is a wide range of sampling methods, none of which can be described as the correct or best method. The most effective method will depend on the requirements of the survey and the characteristics of the sampling units. The most commonly used sample methods in geographical surveys are:

> Random sampling
> Stratification with a uniform sampling fraction
> Systematic sampling from lists
> Line sampling
> Principle of the moving observer

In a random sample the sampling units are selected at random from the population of sampling units. Each sampling unit has an equal chance of being selected and various tables of random numbers can be used to ensure this. There are many variations on this basic sampling method but all focus on random choice.

In stratified samples the population of sampling units is divided into groups or *strata* before sample selection. The strata may contain the same number of units in each, or differing numbers, but from each of the strata the same proportion of units is chosen on a random basis to provide the sample. The advantages of a stratified sample are that it often increases the overall accuracy of the estimates produced by the sample, and secondly it ensures that particular subdivisions of the population, which may be of special interest, are included in the final sample.

In systematic sampling the population of sampling units are listed and then a regular choice of unit is made. It is customary to select the first unit at random, and then successive units are determined by the particular sampling interval chosen. For example, a sample drawn from a list may begin with the sixth unit (chosen at random) and then every subsequent twelfth unit (sampling interval of 12). The sampling interval will be chosen to produce the desired sample size. Sampling from lists, such as the Electoral Register for an adult population sample, can be easier than a random sample but care must be taken to ensure that there is no natural regularity within the list which could bias the result.

Line sampling means taking a sample within a landscape or from a map. This method is widely used in the calculation of land-use areas. Land-use types are defined, then a series of parallel lines are drawn on the map. Land-use changes along each line are noted and percentage areas of the different land uses are calculated according to the proportion of the line which has been drawn across them. Line sampling is also used widely in other aspects of geography, notably in beach and soil studies.

The principle of the moving observer is a method of sampling the number of individuals moving about within a given area and is widely used in surveys of pedestrian activity within cities. A moving observer crosses an area in one direction counting the number of people he passes – in either direction – and deducting the number who pass him. The count is repeated in crossing the area in the opposite direction. The average of the two counts provides an estimate of the number of people in the area. These five sampling methods are commonly used but there are many others available for specialist survey work.

Sample size

The second important decision in undertaking a survey, alongside choice of sample method, is choice of sample size. Generally the larger the sample the more accurate it will be and a sample of fifty is the minimum required for even the simplest of surveys. A potential sample of fewer than fifty will usually mean that it is feasible to carry out a census because the total population will be small. It is the absolute sample size, and not the percentage of the population of sampling units, which is the primary determinant of accuracy. Optimal sample sizes depend on the sample method. As a general rule of thumb the following method gives an approximation to the sample size required for a random sample, assuming a sample error of two per cent.

The first step is to estimate the approximate proportion of sample units which are expected to have the feature being surveyed. For example, in a survey to estimate the number of people visiting a particular shopping centre the percentage (termed p) might be

60%. Then the calculation is made ($p(100-p)$), in this case 60(100-60) = 2400. The sample error tolerated is usually two per cent. The required sample size is then

$$\frac{1.96 \times 2400}{2} = 2352$$

The figure 1.96 is a constant.

For most relatively simple geographical surveys sample sizes of greater than 2500 are rarely needed, and an accuracy of two per cent is generally accepted as adequate. This same equation may be used to calculate sample error for other percentage figures generated by the survey. If the proportion of shoppers was 25% instead of 60%, then the sample error in this survey becomes $\frac{1.96 \times 1875}{2352} = 1.6\%$. Even a survey exercise involving a survey of 2500 sampling units can be a major exercise, but it is often far less arduous than a full census. The use of stratified samples rather than a random sample usually produces greater accuracy for a given sample size, or alternatively stratification can mean that the sample size can be lowered without a reduction in accuracy of the estimates produced.

This general method of calculation of sample size assumes a large population of sampling units. If this population is relatively small then the size of the sample must be amended to

$$\frac{\text{calculated value}}{1 + (\text{calculated value/population of sampling units})}$$

In the above example, if the population of sampling units was only 1000, then the new sample size would be

$$\frac{2352}{1 + (2352/1000)} = \frac{2352}{3.352} = 701$$

This is for a random sample, and again in a stratified sample the sample size could be reduced and an accuracy of plus or minus two per cent retained. If there is a large population of sampling units then again, as a very general rule, a sample of 2400 will give an accuracy or maximum error of ±2%; 400 gives ±5%; and 100 gives ±10%. There is always likely to be a certain level of inaccuracy in small samples, and it is frequently much more satisfactory, when possible, to utilise the results of a survey carried out by a large organisation.

Economic surveys

The examples so far discussed in this chapter have been mainly social surveys of people and their activities, or of households. A second major area of survey in which the geographer is involved is economic activity. Surveys of firms may be undertaken to discover simple data, such as their location, or more complex data, such as their locational decision making activity, the origin of raw materials and destination of finished products, or their employment numbers and types. In the social surveys it is important to distinguish between surveys of individuals and surveys of households. In economic surveys it is important to distinguish between establishments and organisations.

An establishment is a building complex in which activity takes place, for example a factory, shop or school. An organisation is the controlling mechanism of the establishment, for example the manufacturing firm, retailing company or county education authority. An organisation may control many establishments, sometimes even of different kinds, as with major oil organisations who control oil refineries, oil wells, petrol filling stations, storage depots and so on. Alternatively an organisation may operate through a single establishment. Most frequently geographers are interested in establishments in their attempts to *describe* the landscape, but to *explain* the features of the economic landscape surveys of organisations are important.

As with social surveys, various international and national agencies are involved with economic surveys and attempts have been made to develop survey frameworks. In Britain the usual classification of economic activity is based on the *standard industrial classification*, which is an organised list of the different types of economic activity. The total economy is seen as comprising ten *divisions*:

0   Agriculture, forestry and fishing
1   Energy and water supply industries
2   Extraction of minerals and ores other than fuels; manufacture of metals, mineral products and chemicals
3   Metal goods, engineering and vehicles industries
4   Other manufacturing industries
5   Construction

6   Distribution, hotels and catering; repairs
7   Transport and communication
8   Banking, finance, insurance, business services and leasing
9   Other services.

The ten divisions are divided into *classes*, of which there are a total of sixty. Each class is referred to by two numbers: the first is the number of the division, and the second is a class number within a division. For example, class 32 is mechanical engineering and is the second class within division 3; class 93 is education and is the third class within division 9. Classes are further subdivided into industry *groups*, signified by three figures; and groups are again divided into *activities*, which are shown by four figures. In total, 222 groups and 334 activities are recognised. Within division 4, for example, there are eight classes and class 47 is the manufacture of paper and paper products, printing and publishing.

Class 47 has three industry groups:

> 471 pulp paper and board
> 472 conversion of paper and board
> 475 printing and publishing.

Within 475 there are four classified activities:

> 4751 printing and publishing of newspapers
> 4752 printing and publishing of periodicals
> 4753 printing and publishing of books
> 4754 other printing and publishing.

The standard industrial classification provides a basis for the classification of survey results and a framework within which sample surveys can be carried out. Other countries have similar classifications and official surveys and resulting published statistics use the classification. In the USA there is a similar four figure classification to that in the UK. An example from the USA is:

> *Division A* Agriculture, forestry and fisheries
> *Major group 01* Agricultural production
> *Group* 011 Field crops
> *Industry 0112* Cotton

For international comparisons, the United Nations has a classification based on a three figure code, providing 900 individual activity classes. General economic surveys which collect information on items such as location, employment, wages, capital investment, sales, profit, energy use and so on use these standard classifications. They are also used in more specialist surveys of a particular branch of activity.

The standard industrial classification depends on the allocation of establishments (factories, offices etc) to a particular category in the classification. Economic activities may be classified by the jobs which are carried out. A truck driver working in a chemical factory would be classified as a chemical industry worker under the SIC. Alternatively it may be more useful to classify such a worker as a transport worker, along with truck drivers working at mine sites, those employed by transport firms or in many other industries. Such a classification would be an *occupation* based classification as distinct from the *industry* based SIC. Just as there are national and international SICs, there are also standard classifications of occupation. It is important to keep the distinction between occupation and industry clear. A typist is in a specific occupation identified in most occupational classifications, but such a person could work in almost any industry and also be classified in any one of several hundred industry classes. In all geographical survey work it is vital to define the basis of the survey, whether of individuals or households, establishments or organisations, industry or occupation, farms or fields.

Survey analysis

Geographers are concerned with surveys because they provide data about the landscape which are suitable for analysis. Just as geographers, although they are not themselves cartographers or photogrametricians, need to have some knowledge of map making and air photography techniques, so, although they are not survey scientists, they need also to know about survey methods in order to appreciate the accuracy or otherwise of survey data, before meaningful analyses can be undertaken. In order to interpret maps, air photographs and survey results it is necessary to have some understanding of how the material was collected and the problems associated with the sources. Maps and photographs have information on the edge. In survey reports there is a written section on methods, classification systems, and often a copy of the questionnaire which has been used.

This section may only be in small print, tucked away at the end of the report, but it is just as important as a map key.

Again, as in map interpretation and air photography analysis, computer techniques are having a considerable impact on survey methods and analysis. One feature of a computer is that it can carry out repetitive tasks very quickly and with a high level of accuracy. Much analysis of survey data is simply counting up the number of responses of a particular type and this task is easily carried out by computers, as sorting and classifying large numbers of responses is a task for which they are particularly appropriate. Computer methods have also speeded up the analysis of major surveys such as national population censuses. Geographers can also extract a greater quantity of information from a particular survey through the use of more sophisticated analytical methods. If a computer is linked to map making equipment, raw survey data can be fed in, the survey analysed and the results presented in map form drawn by the computer. In such an exercise the geographer must know not only how to analyse the survey and draw maps, but also how to tell the computer to carry out survey analysis and map drawing.

PART THREE

Patterns in the Landscape

7

World Distributions

The discussion in earlier chapters about landscape and the ways landscapes change suggested an approach to geography which focuses on landscape as we see it through the window or from a hilltop. Such description and analysis of relatively small areas makes it easy to forget that landscape variety also occurs on a much larger scale. On a world scale major landscape regions can be defined using a relatively small number of key features and many of the world distributions of landscape features show how closely the various aspects which comprise landscape are inter-linked. Before looking in the next chapter at some of these associations on a world scale, it is worthwhile considering some important and fundamental world distributions.

Patterns of world climate

Climate is one of the most potent landscape factors, influencing both physical and human processes of landscape change. A climate is a particular combination of meteorological variables, such as temperature, moisture, winds and precipitation, which exists at a specific place. Average values and extremes of these measures are used to define individual climates and the values are based on long periods, often several decades, of observation. The pattern of seasonal change in these variables is an important aspect of the description of climate. Many attempts have been made to classify climates and to produce world maps. Most of the older atlases have a map of world climates and many use classifications based on one devised by Köppen. This classification, with minor modifications, defines six major climatic types. Each major type may contain subdivisions related to particular precipitation and temperature conditions. Each type and subtype is identified by a sequence of up to three letters. The first letter defines to which of the six major types the climate belongs, and subsequent letters define the particular precipitation and temperature conditions. The six major types are:

A   Tropical climates. All monthly temperatures over 18°C
B   Dry climates
C   Temperate climates. Mean temperature of the coldest month is between 18°C and −3°C
D   Snow climates. Mean temperature of the warmest month over 10°C and of the coldest under −3°C
E   Ice climates. Mean temperature of the warmest month under 10°C
H   Highland climates.

A second letter may be added to describe specific precipitation conditions as follows:

s   steppe climate
w   desert climate
f   sufficient precipitation in all months
m   monsoon cycle climate
n   dry season in summer in respective hemisphere
p   dry season in winter in respective hemisphere.

A third letter is used to supplement the temperature based divisions as follows:

a   warmest month mean over 22°C
b   warmest month mean under 22°C and at least four months have means over 10°C
c   fewer than four months with means over 10°C
d   fewer than four months with means over 10°C and coldest month mean less than −38°C
h   mean annual temperature over 18°C
k   mean annual temperature under 18°C

Using this scheme the British Isles is classified as *Cfb* – a temperate climate without precipitation or temperature extremes. The European coast of the Mediterranean is classified as *Cna*; the equatorial areas of Africa and Latin America as *Af*, and Saharan Africa as *Bph*.

Group *A* climates comprise three main subtypes: *Af*, which are hot and wet with a steady supply of rain; *Am*, which are monsoon climates with a marked very wet period; and *Ap*, which has extremes of temperature and clearly defined wet and dry seasons. All these class *A* climates are within the tropics, with the high sun season bringing rain and the low sun bringing drought in the *Ap* climate.

Type *B* climates comprise two major types, although there are a number of other dry climates in relatively small areas. The two main types are *Bw* and *Bsh*. With *Bw* climates rainfall is low, often less than 100 mm per year, and temperature extremes are considerable in both hot deserts, *Bwh*, and the colder ones, *Bwk*. The desert margins have a climate with sufficient precipitation (up to approximately 500 mm) to sustain hardy, but sparse, vegetation and again these steppe climates may be subdivided by temperature into *Bsh* and *Bsk*.

Type *C* climates are mid-latitude climates without severe or extreme conditions, although occasional winter conditions can be cold and summer days may be quite hot. There are several varieties of type *C*, conditioned by the time of the wet season and the temperatures of the warmest month.

Snow climates of type *D* are northern hemisphere continental climates with cold snowy winters and short cool summers. Generally there is a progression from south to north of *Dfa*, *Dfb*, *Dfc*. In North East Asia there are areas of winter dryness giving climates of *Dwa*, *Dwb* and *Dwc*. Type *E* climates have no more than four months of the year with mean temperatures above 0°C and the warmest month is less than 10°C. Highland climates, type *H*, occur where altitude influences climate more than any other single variable. The overriding consideration is the coldness of the air, and frequently precipitation is relatively slight.

This scheme allows climates to be described, but tells us little about why particular climates occur in particular places or what causes climates to differ one from another. Maps of the distribution of these climatic types, or very similar ones, are provided in most atlases.

Climatic types based on air masses

An alternative way of viewing world climate is in terms of air masses. An air mass is simply a large body of air of considerable horizontal extent, covering thousands of square kilometres, characterised by relative uniformity of temperature and humidity at a particular altitude level. Air masses can be of a large vertical extent (3–6 km), or in some cases may be shallower, with another air mass lying above the ground level one. Air masses have distinct boundaries, termed fronts, between adjacent masses and these interfaces are often zones of rapidly changing weather conditions (see Chapter 3). An air mass assumes its temperature and humidity characteristics from its source region. For example, air accumulating over a cold snowcovered landmass will lose heat by radiation and become cold, and being cold the moisture content will be low. Alternatively an air mass formed over a tropical ocean will be relatively warm and will hold water from sea surface evaporation, giving a high moisture content. Air masses are classified according to the latitude of the source area (influencing the temperature) and the type of source area, whether land or water (influencing humidity). Four basic types of air mass are recognised according to their source by temperature. These are:

A (or *AA*)	*Arctic* formed over the Arctic (or Antarctic) areas
P	*Polar* formed between latitudes 50° and 65° N and S
T	*Tropical* formed in the sub-tropical latitudes 20° to 35° N and S
E	*Equatorial* formed in areas close to the equator.

A and *P* masses are cold; *T* and *E* are warm. Each of these four types may be formed over land or water and a second letter is then added to describe a particular mass in terms of moisture content:

m	*maritime* air masses are formed over oceans and are relatively moist
c	*continental* air masses are formed over large landmasses and are relatively dry

Air masses can be defined, for example as *cP* – continental polar – or *mT* – maritime tropical. Additionally the letters *w* and *k* may be added to describe the temperature of the air mass relative to that of the area over which it lies. The letter *k* means the air mass is colder than the surface below and *w* means it is warmer. An air mass designated *cPk* is continental polar air which is colder than the surface over which it lies at the time.

The world pattern of climates becomes easier to understand when analysed in respect of air masses. Three groups of climates, *A*, *B* and *C* can be identified in terms of air masses and these, with their sub-groups, 1–14, are shown in Fig. 7.1. Group *A* climates are dominated by tropical and equatorial air masses. Equatorial climates (subgroup 1) are formed where warm moist *mT* and *mE* air masses give heavy rainfall in convectional storms. Easterly winds in the tropics bring moist *mT* air masses into contact with eastern coasts of continents, producing areas of heavy rainfall and even temperatures, but rainfall is seasonal due to shifts of air masses associated with the overhead sun (type 2). Desert climates (type 3) are the source of *cT* air masses with subsiding dry air. On west coasts there are relatively cool and extremely dry climates caused by subsiding *mT* air (type 4). Only in Group *A* are there tropical wet–dry climates, where the seasonal alternation between *mT* or *mE* air masses and dry *cT* gives a climate with a wet season associated with the high sun.

Group *C* climates comprise cold climates dominated by polar, arctic and antarctic air masses. These are climate types 11–14 in Fig. 7.1. Areas of continental subarctic climate coincide with the source areas of *cP* air masses which in winter are stable and very cold. Occasional incursions of *mP* air supply light precipitation. Marine subarctic climates (type 13) are more influenced by *mP* air masses, whilst the tundra climates occur along the front between arctic and polar air masses and cyclonic storms are frequent. The very cold stable *A* and *AA* air masses produce the icecap climates of Greenland and Antarctica.

Group *B* climates occur in the mid-latitude regions where both tropical and polar air masses interact. Cyclonic storms travelling from east to west bring alternately, often rapidly, tropical and polar air masses over a region to produce a great variety of weather both daily and seasonally. The humid subtropical climates (type 6) occur on eastern continental margins where *mT* air masses move from the western edges of their source area. During the period of the high sun season rainfall is abundant and temperatures high. In winter, *cP* air

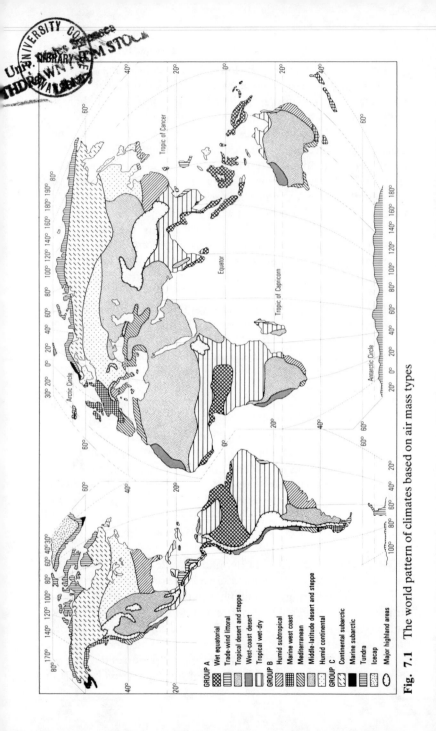

GROUP A
- Wet equatorial
- Trade-wind littoral
- Tropical desert and steppe
- West-coast desert
- Tropical wet-dry

GROUP B
- Humid subtropical
- Marine west coast
- Mediterranean
- Middle-latitude desert and steppe
- Humid continental

GROUP C
- Continental subarctic
- Marine subarctic
- Tundra
- Icecap
- Major highland areas

Fig. 7.1 The world pattern of climates based on air mass types

masses become important and cyclonic activity occurs along the *cP–mT* front. On west coasts (type 7) of continents in the 40–60° North and South latitudes, cool moist *mP* air is important, with cyclonic activity especially in winter bringing increased cloud and precipitation. West coast climates at 30–45° North and South (type 8) show a wet winter and dry summer pattern caused by *mP* in winter with associated cyclonic storms and *mT* in summer bringing warm dry air. The interaction between polar and tropical air masses also occurs in central and eastern continental areas at mid latitudes and gives rise to the humid continental climates (type 10). Seasonal contrasts are strong and summer rain comes from invading *mT* air masses and winter cold from *cP* northern source areas. Continental interiors where *cT* and *cP* air masses dominate in summer and winter respectively have dry climates with extreme seasonal ranges (type 9) and produce the mid-latitude desert and steppe regions. The particular dominance of an air mass at any one period in the mid-latitudes is determined by the westerly winds which flow in the mid troposphere in a wavelike pattern of ridges and troughs superimposing alternate north–south air flows on an overall dominant west to east air flow. Where the dominant westerlies have a component blowing from north to south, cold air masses extend southwards. On the other limb of the wave, where the dominant westerlies have a northward blowing component, warm air masses are funnelled northwards. The variable behaviour of mid-latitude weather is further complicated because the wave pattern of the westerlies varies over time and a particular pattern may last from a few days to several weeks.

The types of climate shown in Fig. 7.1 are broadly similar to those described by classifications such as Köppen's, but the understanding of air masses allows some explanation of the 'where and why' of world climates. The characteristics of particular air masses themselves are important but often the interaction zone between adjacent air masses creates atmospheric processes of considerable strength and regions in this frontal belt are subject to rapid changes in weather. The polar frontal zone which separates polar (*mP* or *cP*) and maritime tropical masses is a zone of intense activity because of the considerable difference between the interacting air masses. The front in the equatorial belt, the intertropical convergence, is not as clearly defined because the air masses on each side are less distinctive. Nonetheless occasional severe tropical cyclonic activity can occur and typhoons or hurricanes may be formed. Source areas for such summer storms tend to be at approximately 10° North or South, with storms travelling westwards and subsequently swinging northwards in the northern hemisphere and southward south of the equator. Although relatively rare events, these storms can cause very considerable destruction to both natural and man-made landscapes through high winds, heavy rain and high tides.

Patterns of world vegetation

The pattern of world climate is an important basis for understanding many other patterns of world landscape components. Soils, vegetation and agriculture, themselves closely interrelated, are directly influenced by climate. The world patterns of population distribution and wealth are influenced indirectly by climate. Most atlases contain a map of world vegetation types, or *biomes*, which show a sequence of forest, grassland and desert vegetation which may be related to climatic conditions. A distribution map of the major soil groups shows a pattern of *zonal soils* which is related to climate and vegetation. Such maps tend to suggest a rapid transition from one vegetation and soil type to another, but in reality the zone of change is often broad and rather indistinct. They also suggest uniformity within a biome, and this is an oversimplification. In fact, the complex interaction of such factors as soils, altitude, slope and drainage, aspect and local climate produce a differing climax vegetation in localised areas within a biome.

In equatorial rain forests where there is little variation in climate and the growing season is continuous, hardwood evergreen trees are dominant and a vegetation canopy is more or less continuous at about 30 m. Tree foliage tends to be concentrated in the crowns, and this canopy contains a variety of lianas, epiphytes and parasites. Only occasionally do taller trees rise above the canopy, whilst under it there are young trees, again with epiphytes and parasites, and only a sparse ground cover of shade tolerant or shade demanding species. Species variety is very high. In seasonal forests, both tropical and temperate, species diversity is lower and there are frequent stands of a single species where local climate, soil and site characteristics produce optimal conditions for a particular plant. In such places a continuous even canopy of foliage develops, but where species are mixed taller species rise above a rather uneven and sometimes broken canopy. As the length of the dry season increases, so the canopy becomes more broken. The density of the canopy influences the amount of ground cover and low shrub development. The densest forests of true 'jungle' conditions are found in tropical seasonal forests such as the monsoon forests of South East Asia, where the dry season is long enough to thin the canopy and allow sunlight to reach the ground but not long enough to produce drought, so that a dense undergrowth can still develop. In the northern coniferous forests of Asia and North America the growing season lasts only from June to August and a dense tree canopy occurs, composed of a very few species of softwoods such as spruce and fir. The needle litter is slow to decompose because of the low temperatures, the trees are often growing on nutrient sparse soils, and the ground litter is highly acidic. These factors together with the dense shade inhibit the growth of ground cover.

Woodlands have a discontinuous canopy of evergreen or deciduous species in clumps of two or three trees. Since the trees grow in full sunlight foliage exists from the tree crown to the shrub level. Young trees tend to grow away from adult trees, unlike the forest types where the young grow in the shade of the adults. The ground cover in woodlands is of shrubs and herbs, which also may grow in clumps. Tree foliage close to the ground, together with shrub and herb cover, provide a rich food supply for insects and animals, particularly grazing animals. In woodlands an individual species of tree will probably show a clustered spatial pattern, whilst in rain forests spatial patterns are more random.

In grasslands, herbs and shrubs are widely present. Just as forest vegetation has several levels, or *storeys*, so also are grasslands multistoreyed. There is often more than one storey of grasses, herbs or shrubs, and also a very low ground cover of plants. Most so-called natural grassland has been used by man and changed by him through burning and ploughing. *Savanna* lands are tropical grasslands with scattered woody plants and may be seen as an intermediate vegetation type between very broken woodland and grasslands. These savanna lands and some grasslands such as the North American prairies may be the result of repeated firing of woodlands, not only by human but also from natural causes. The

presence of grazing animals is probably significant in the main-
tenance of grasslands and the spatial pattern of tree, shrub and grass
species is also related to factors such as slope and soil.

The savanna and grasslands merge into hot, dry and semidesert
lands (see Chapter 8). The high latitude tundra areas of the northern hemisphere are also vast treeless areas, but they experience low
temperatures all year and are widely underlain by permafrost,
which is a condition of permanently frozen subsoil. Vegetation
consists only of ground cover plants such as lichens, dwarf flowering
grass and woody plants that can make use of a growing season as
short as sixty days and can withstand long periods of snow cover.
Much of the animal population consists of summer migrants. Plant
growth is slow, species diversity is low, soils are poorly developed so
that the tundra biome is very fragile, recovering only slowly from
any damage. The tundra and hot deserts are the least productive
regions of the world.

Patterns of vegetation productivity

Different vegetation types produce different amounts of vegetation
material. Vegetation productivity is the amount of trunks, stems
and leaves produced from a given area, and this measure varies in
different climatic regimes. Various measures and indices of vegetation productivity have been calculated, and there are world maps
which indicate the potential usefulness of the regions. Fig. 7.2 is
typical, and whilst other schemes and maps differ in detail, this one
by Paterson provides the general pattern common to all schemes.
The index used by Paterson is:

$$I = \frac{TPGS}{120 \cdot R}$$

where I is the index of productivity

T is the average temperature of the warmest month (°C)

P is precipitation (mm)

G is growing season (months) with average monthly temperature of 3°C or higher

S is solar radiation expressed as a proportion of that at the
poles

R is the annual range of temperature between the average of
the coldest and warmest months (°C).

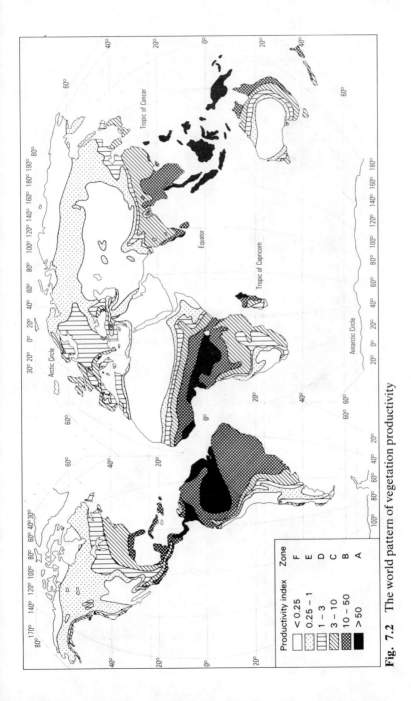

Fig. 7.2 The world pattern of vegetation productivity

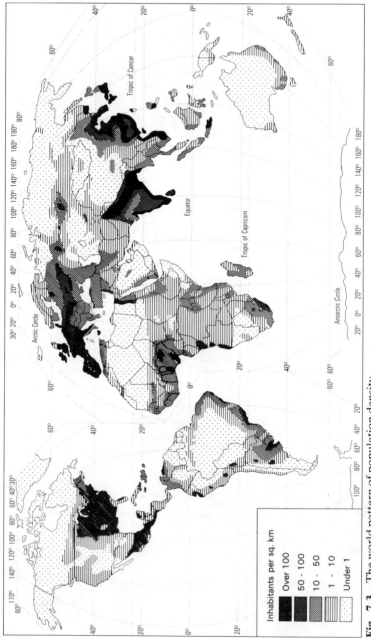

Fig. 7.3 The world pattern of population density

Inhabitants per sq. km

- Over 100
- 50 – 100
- 10 – 50
- 1 – 10
- Under 1

Values of this index are very low (less than 0.25) in the low productivity polar and desert areas and may rise to over 100 in the equatorial regions. Fig. 7.2 shows six zones. Around the virtually nonproductive zones is a zone of low productivity which is an extensive area in Asia and North America but is relatively small around the hot deserts. The areas with cool temperate climates tend to have productivity values between 1 and 3. Zone *C* is of medium productivity and includes some of the densely populated agricultural regions of the world. The high productivity zones *A* and *B* are restricted to the tropics. The index shows potential productivity; many factors either inhibit areas from achieving their potential or enhance the productive capacity of a particular region. For example, plant growth may be reduced by pollution but be increased by irrigation. Fig. 7.2 is a broad zonal map and within each zone there will be a considerable range of index scores of potential productivity resulting from particular local conditions and also a great range of actual productivity levels resulting from different degrees of human interference with the processes of plant and tree growth.

World population patterns

The distribution of people in the world is as fundamental a geographical distribution as that of climate. Earlier chapters have shown how at a local level man has modified the landscape and environment. The totalling of these local modifications, merely through living in an area, can result in major, global, changes in environments. We have seen the importance of population censuses in finding out how many people there are and where they live. Many countries do not have censuses, so maps such as Fig. 7.3 showing the density of population are based on estimated numbers and distribution. At a general world scale these estimates are acceptable, but they can create problems in more detailed studies. World population increased from 2044 million in 1930 to 2486 million in 1930 to 2486 million in 1950, but during the last three decades much more rapid growth has occurred. By 1970 population numbered 3621 million, and estimates for 2000 suggest 6000 million. Fig. 7.3 refers to the late 1970s, when the world population was approximately 4200 million people.

A very considerable range of population densities is seen on Fig. 7.3. Vast areas of the world are virtually empty, and seventy per cent of the population live on less than ten per cent of the land area. Four concentrations of population are outstanding. The first is South Asia, including India, where high densities occur over a large area. Second is East Asia, including eastern China, Japan and Vietnam. The third is Europe, where very high densities are less continuous but the general level of population density remains above fifty people per square kilometre. The fourth, the smallest concentration, occurs in North East USA, where a number of small areas of high density have become joined together to produce one high density concentration of over 120 million people. This is a small total, however, compared with the 700 million of India and Bangladesh. There are several smaller areas of high density, including parts of Indonesia, the world's fifth largest country in total population terms; Nigeria; the lower Nile valley in Egypt; coastal Brazil; and the Plate estuary area of Argentina/Uruguay.

Earlier this century geographers tried to explain the distribution of population simply as a response to climate, both directly in controlling man's activities in areas with a comfortable climate and indirectly through its effect on food supply. These views have been discarded and more complex reasons for the pattern of population are now being sought. Climate may be one of many influences on the distribution, but it is not the determining one. Man relates to his environment through culture. In creating a culture man responds more positively than other living organisms to the natural environment. He can make artefacts, develop techniques for using them, and communicate his experiences to others: in this way he can modify the natural environment to suit his own purpose. The world distribution on a broad scale is a response to broad cultural change and economic advance. The growth of Europe as a major population centre owes much to the cultural changes of the agricultural and industrial revolutions since the eighteenth century. The emergent area of high population density in California can be related to cultural changes and new technologies of the mid-twentieth century, whilst the population concentrations in the fertile river valleys in Asia date back several thousand years to social and technological changes in the use of water. In many of these areas population remains rural based, although some of the earliest cities are found there, unlike the population concentrations in North America and Europe which depend on modern urban society. The domination of rural areas in the population distribution in Asia is changing as major modern cities emerge and culture is again transformed.

Unlike the world distribution of climate and vegetation discussed in this chapter, the pattern of population location and density is changing significantly, both as population numbers increase, and also as major migrations of population take place (see Chapter 8). The European movement to North America in recent historic time illustrates how the map of population distributions can be changed. More recent still are the changes due to the rapid growth of the world population in the last forty years. The growth of population has served to extend the range of population densities and to initiate new small regions of high density. Density levels have increased even higher in already well populated areas in Asia, but also new high density nodes have appeared. The number of cities with over one million population has increased from eighty in the 1950s to almost 200 by the 1980s. The number of even larger cities of at least five million population has increased from six to 1950 to approximately thirty in 1980, and by 2000 it is estimated that there will be a further twenty of these giant cities. Judging by relatively modest estimates, Mexico City may by that time have 25 million people.

World patterns of wealth

Underlying the map in Fig. 7.3 are many aspects of the variety of population. Population change, age, sex, level of literacy, religion, incidence of disease, and so on can all be shown at a world scale and some of these are explored in later chapters. For the moment it is useful to look at a fourth very important basic world pattern alongside those of climate, vegetation and population – that of wealth. As with the other variables, wealth is difficult to map with a single measure and every attempt to produce a world distribution map can be criticised in its detail. A commonly used measure of wealth is *gross domestic product* per person.

Gross domestic product (GDP) is the total value of all economic activities within the country. GDP per person or per capita provides an index for the comparison of wealth amongst countries, but it is important to remember that it does not tell us anything about how the wealth is distributed within a country. The wealth may be evenly spread through the population or it might be concentrated into a relatively small number of wealthy people and the majority may be very poor. It must be remembered that there are limitations to the use of GDP per capita, including many problems of actually calculating GDP.

Fig. 7.4 shows the world pattern of GDP per capita in 1977 in US dollars. The range of values is from 80 in Bhutan to 12270 in Kuwait, both countries having just over a million people. The world pattern is striking but predictable. Low income countries may be defined as those with a GDP per capita of $300 or less. Included here are several African countries, but also, significantly, India, Pakistan, Bangladesh, Vietnam and Indonesia, which together have almost 1000 million people. Middle income countries can be defined as those with GDP per capita between $300 and $3000 but it is useful, as in Fig. 7.4, to divide them into two groups. Again amongst these countries are many with massive populations, including China with a GDP per capita of $390. Amongst the richer countries with GDP per capita of over $3000, only three have GDP per capita of over $9000 – Sweden, Switzerland and Kuwait – but in general these richer countries are the industrialised countries of Europe, together with other industrially-based countries such as the USA, Japan and Australia.

GDP per capita changes, and the map of 1977 would have some notable differences if compared with one of 1950. Not only have the oil exporting countries in the Middle East shown a considerable increase in wealth, but two other important changes have occurred. First the range of values has increased considerably, so that the rich have grown richer at a faster rate than the very poor; secondly, some middle income countries have increased their wealth faster than many richer countries. Table 7.1 shows the growth rate per year (if prices are fixed at 1975 levels) in GDP per capita for 1960–1970 and 1970–1980. The very poor countries have an extremely low growth rate, with individual countries such as Chad, Somalia, Niger and Bangladesh showing declines in wealth. The oil exporting countries have high rates of increase and industrial countries have rates of increase of between two and four per cent per year. Several countries in Southern Europe – Spain, Greece and Portugal, for example – and some in Asia – South Korea, Hong Kong, and Singapore – have growth rates of between six and eight per cent per

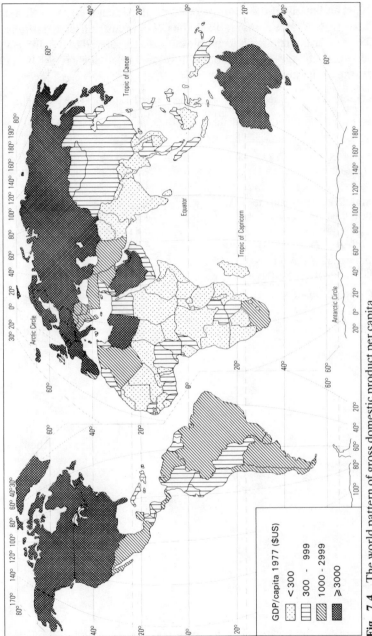

Fig. 7.4 The world pattern of gross domestic product per capita

GDP/capita 1977 ($US)

- < 300
- 300 - 999
- 1000 - 2999
- ≥ 3000

year, and many of the countries with GDP per capita of over $1000 have growth rates comparable to the richer industrialised countries. The pattern of change is a complex one and is related to the process of social change and economic development, which is the main theme of the next chapter.

Table 7.1   Average annual percentage growth rates of GDP per capita by groups of countries

	1960–1970	*1970–1980*
Low income countries – All	1.8	1.7
in Africa	1.5	0.2
in Asia	1.8	2.0
Middle income countries – All	3.9	2.9
in East Asia	4.9	6.2
in South Europe	5.8	3.4
in Latin America	2.9	2.6
Industrialised countries	3.8	2.7
Oil exporting countries	8.2	8.1

(Source: *United Nations Yearbooks*)

8

Associations Amongst Distributions

Measures of economic development

The world map of gross domestic product per person described in the last chapter was considered as a map of relative levels of economic development. Many other measures of economic and social development are available. They can be mapped, and it is possible to produce atlases devoted to these measures to try to obtain a comprehensive view of 'development'. Table 8.1 gives a list of the sorts of measure that are used to study different aspects of economic and social development. Each of these measures may be considered either individually or together, measuring essentially the same idea. In many cases the measures are interrelated, and if maps are drawn then the resulting distribution can appear quite similar. So, for example, a world map of life expectancy at birth, whilst showing considerable differences from seventy-six years for Japan to only thirty-nine in Ethiopia, has similar broad patterns of variation to a map of adult literacy rate, which varies from almost a hundred per cent in Europe, North America and Japan to only ten per cent in Ethiopia. If other measures from Table 8.1 are mapped by country, an equally wide range of values emerges and a sharp contrast is present between high and low income countries. The distributions of the measures are strongly associated with one another. The complex idea of development can be measured generally by any one indicator, but it can be measured more specifically by using several indicators together.

Various terms have been used to signify the poor countries of the

Table 8.1 Selected indicators of economic development

	More developed countries	*Less developed countries*
Gross national product per capita	high	low
Death rate	low	high
Birth rate	low	high
Infant mortality rate	low	high
Doctors per person	high	low
Proportion of the population who are literate	high	low
Railway density	high	low
Newspaper circulation per person	high	low
Telephones per person	high	low
Motor vehicles per person	high	low
Foreign trade per person	high	low
Proportion of the population employed in agriculture	low	high
Capital per hectare of agricultural land	high	low
Agricultural yields per farm worker	high	low
Electricity consumption per capita	high	low
Oil consumption per capita	high	low

world. 'Backward countries', 'undeveloped', 'underdeveloped', then later 'developing' were commonly used terms. Now, 'less developed' countries (LDCs), 'Third World', or even 'so-called Third World' countries are terms in vogue to describe this group of countries. The common characteristic of all these countries is that they are poor and tend to have been left behind as economic growth has taken place in Western Europe and North America (the First World countries) and in the communist countries (Second World). What exactly constitutes a Third World country is difficult to define, for whilst there would be general agreement that countries such as Pakistan, El Salvador and Tanzania belong to the group, it is difficult to decide about countries such as Turkey, Malaysia and Mexico. Some of the indicators would suggest that they are developing rapidly and may join the developed countries in the next decade;

other indicators are more depressing and suggest these countries are being left behind by growth in the countries of the First World. In Malaysia, for example, per capita food production increased by thirteen per cent between 1970 and 1975, life expectancy in the late 1970s was sixty-seven, but GNP per capita was only $930 US, which is about one tenth of that of Sweden or Switzerland.

Despite some variation in the middle income countries, the less developed countries do have a number of common characteristics in addition to poverty. Generally the level of commercialisation of agriculture is low, with many of the population producing much of their own food. Agricultural technology is noticeable by its absence and manufacturing industry is minimal. The commercial aspects of agriculture show a heavy dependence on a few crops. Maybe only one or two are grown, such as coffee, cocoa, cotton or rubber, and exported to developed countries. Any major mineral production often goes the same way, so that the economy of these countries is often at the mercy of fluctuating world commodity prices and terms of trade. Common population characteristics also typify the Third World. Families tend to be large, birth rates are high, life expectancy is low, so there is a large proportion of young non-productive people in the population. In some countries almost fifty per cent of the population is under fifteen, which by First World standards would be considered below the age when someone can enter the productive workforce, but in the Third World many young people under fifteen work full-time. The increase in the proportion of young people in less developed countries results largely from the diffusion to the Third World of great improvements in medicine and public health services, particularly since 1945. This has led to a decline in death rates without a comparable decline in birth rates. Natural increase is frequently above 1.8 per cent in the developing countries, and in some cases is over 4 per cent where the decline in mortality rates has been largest, as for example in Middle America. Some death rates have fallen to below ten per 1000, which is lower than for many of the countries of Western Europe. In the case of small countries such as Trinidad, Singapore and Mauritius, low death rates are commonplace and public health programmes can reduce death rates to as low as five per 1000. The population age–sex pyramids, as shown in Fig. 8.1, of the Third World show a very broad base and tapering shape with relatively few old people.

In developed countries such diagrams are quite a different shape and show broad equality in the proportions of people in age groups up to sixty.

In the 1980s we might expect to see a change in these patterns as more Third World countries adopt family planning policies and birth rates begin to fall. In 1960 only three less developed countries had effective birth control policies, but by 1971 sixty more had adopted such schemes. Although policies are effective to different degrees, the cumulative effect in the 1980s will be that of reducing birth rates. Birth control policies on their own have some effect, but there also has to be a change in the wishes and aspirations of the population if birth rates are to decline drastically. In many countries there is growing evidence that women now want fewer children and so are willing to accept family planning and birth control techniques. In South Korea the birth rate fell from 34 to 24 per 1000 in a decade because of later marriage and the acceptance by the population of contraceptive practices. Birth control is not accepted in all countries, for example, in Latin America, where the influence of the Roman Catholic Church is strong. Religion is only one likely influence, and the prestige of male virility is often important. Furthermore family planning is sometimes criticised as an attempt by American and European nations to impose their values on the Third World. Some countries, particularly in Latin America, still regard themselves as underpopulated, having large areas available for exploitation by future generations. Not all Third World countries accept the view that population numbers must be limited.

Yet another common problem of the Third World countries is that of hunger. Paul and Anne Ehrlich in their book *Population, Resources, Environment*, published over a decade ago in 1970, wrote, 'Look for a moment at the situation in those nations that most of us prefer to label with the euphemism "under-developed", but which might just as accurately be described as "hungry"'. Average daily per capita calorie supply, which in Western Europe is about 3500, falls below 2000 in the poorer parts of the Third World, and with such an average many individual's intakes are well below 1500. If a value of the daily per capita calorie requirement is calculated which takes into account the age and sex distribution, average body weights and environmental temperature, then in much of the Third World daily calorie intake is ten to twenty per

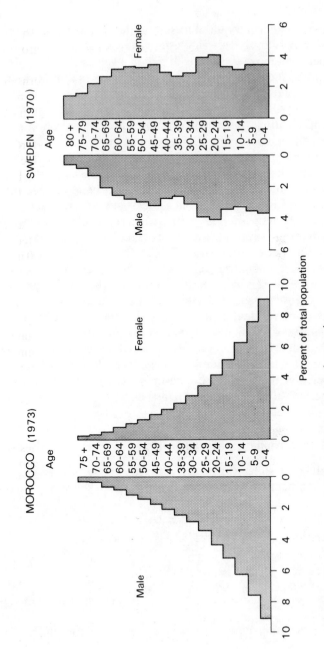

Fig. 8.1 Population age–sex pyramids for selected countries

cent below the level required to sustain the population at normal levels of activity and health. In Western Europe values are twenty to thirty per cent above this norm. If national values of this measure are mapped, then again the same pattern emerges as is seen with measures of wealth, population character and health.

Other measures shown in Table 8.1 give similar results and similar distributional patterns. Transport and communication facilities per person, consumption of metals per person, and consumption of energy supply per person, all show the same dichotomy between developed and developing nations. The idea of 'development' is therefore a very complex one which can be considered from several viewpoints. Different measures show different facets of development, and different pairs of measures are associated in slightly different ways. A scattergram of GDP per capita and energy consumption per capita is shown in Fig. 8.2, and it can be clearly seen that wealthy countries have high values of energy consumption per capita. How closely associated are the two sets of values? Are they more closely associated than GDP per capita and percentage of the population resident in urban areas? In order to understand the world pattern of economic and social development it is important to know the extent of association between the different world distributions which together describe development. To discover the amount of association between these measures statistics may be used to assess quantitatively the amount of association.

A statistical measure of association

A widely-used statistical measure in geography is the Pearson product moment correlation coefficient, usually called 'r'. Values for 'r' vary between 1.0, which signifies that two patterns are completely associated, to zero, which signifies no association. The value also has a sign: plus (positive) means that high values on one measure are associated with high values on the second, and minus (negative) means that high values on one measure are associated with low values on the second. The r value for Fig. 8.2 is 0.86, which shows that the two measures are closely related but are not identical.

The correlation coefficient r can be used to measure the associ-

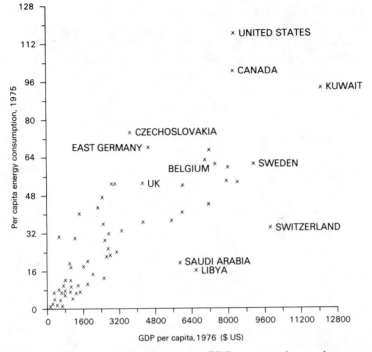

Fig. 8.2 The association between GDP per capita and energy consumption per capita

ation between pairs of measures of other geographical distributions. Values of r are calculated in the following way for any set of pairs of values. Measure 1 is termed X, measure 2 is Y, and the number of pairs of these values we can call N, then:

$$r =$$

$$\frac{N\Sigma xy - (\Sigma x)(\Sigma y)}{[N\Sigma x^2 - (\Sigma x)^2][N\Sigma y^2 - (\Sigma y)^2]}$$

where Σ means 'the sum of'.

So to find a value of r from this equation only six values are needed.

1   N, the total number of observations
2   the sum of the Xs (Σx)
3   the sum of the Ys (Σy)
4   the sum of the squares of x (Σx^2)
5   the sum of the squares of y (Σy^2)
6   the sum of the product of x and y (Σxy)

For ease of calculation a simple table can be set up as follows:

Observations	x	y	x^2	y^2	xy
1	—	—	—	—	—
2	—	—	—	—	—
3	—	—	—	—	—
.	.	.	.	.	.
.	.	.	.	.	.
.	.	.	.	.	.
N	—	—	—	—	—
Totals	Σx	Σy	Σx^2	Σy^2	Σxy

From the totals and the value of N the Pearson product moment correlation coefficient can be calculated easily. Some hand-held calculators provide facilities for the direct calculation of r.

The various measures of development each have their own detailed world distributional pattern and the associations between these measures may be calculated statistically and then analysed. More complex statistical techniques can be used to try and distil the common features from the whole series of correlation scores, thus creating a new variable which effectively describes a more complicated idea than any one of the individual measures. Such an exercise was carried out for the development concept by an American geographer, Brian Berry, who suggested that the forty-three separate measures he used to assess relative economic development in ninety-five countries could be reduced to four. The two most important *factors* identified by this analysis represented the technological development of a nation and its overall demographic characteristics. The technology factor provided an indication of the degree of technological complexity present in each nation's economy. Thus the heavily industrialised countries of the First World have values at one end of this factor, and the low technology agricultural countries of the Third World lie at the other extreme. The second major new factor isolated was a complex demographic measure which emerged from associations amongst individual measures such as birth rate, death rate, infant mortality, urban population, population growth rate, and so on. This second factor integrated together several social variables of development, with extreme values shown for countries with a high level of social welfare such as New Zealand, Norway, Denmark and Sweden, while at the other extreme are densely populated countries such as India, Indonesia, Nigeria and Pakistan, where social services are poorly developed or absent.

Types of agriculture

Similar methods may be used to study the associations amongst the measures that make up other complex ideas. Attempts to define world agricultural regions often involve the use of several criteria in combination to produce a single world distribution map. For example, Whittlesey produced a map of world agriculture as long ago as 1936 using five criteria:

1 Crop and livestock combination
2 Intensity of land use
3 The processing and disposal of products
4 The tools and methods used in farming
5 The complex of structures associated with farm enterprises.

More recently Thoman has produced an alternative world picture by combining the three criteria of:

1 Type of crop and livestock or combination of crops and livestock
2 The intensity of land use
3 Degree of commercialisation.

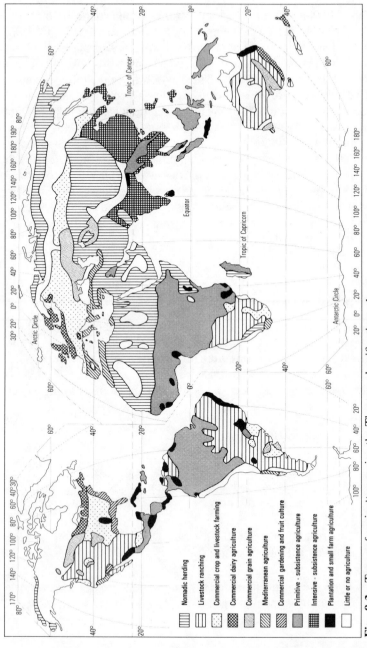

Fig. 8.3 Types of agriculture using the Thoman classification scheme

Legend:

- Nomadic herding
- Livestock ranching
- Commercial crop and livestock farming
- Commercial dairy agriculture
- Commercial grain agriculture
- Mediterranean agriculture
- Commercial gardening and fruit culture
- Primitive - subsistence agriculture
- Intensive - subsistence agriculture
- Plantation and small farm agriculture
- Little or no agriculture

The result of the exercise is shown in Fig. 8.3, which distinguishes ten types of agriculture. Attempts by other geographers using alternative mixes of criteria result in only slightly different world patterns.

An important feature of Fig. 8.3 is the dichotomy between commercial farming and subsistence agriculture. At one extreme is the totally commercial farm, where produce is sold off the farm and the inputs into agriculture such as fertiliser, animal foods and so on are purchased from outside dealers. In subsistence agriculture little is bought in and production is geared to family or village food consumption. For many years this distinction was clearly apparent in parts of Africa and Asia, where peasant subsistence agriculture and commercial plantations existed side by side. In 1950 it was estimated that 1300 million people lived in selfsufficient agricultural economies. Increasingly over the last thirty years subsistence agriculture has given way to commercial enterprise, so that the decline of subsistence economies is a notable feature of the changing society in large parts of the tropical world. The area shown to have subsistence agriculture in Fig. 8.3 reflects the date of the work carried out by Thoman, and gives a good indication of the position in about 1960. Since then a considerable contraction has taken place and many of the subsistence areas contain enclaves of commercial agriculture. Subsistence agriculture is also being reduced through increased use of fertilisers and new plant varieties which are brought into the system. The surplus production resulting from these agricultural improvements is sold out of the system. It is likely that in a few years all agriculture will be commercial to some degree with virtually no areas of totally subsistence agriculture left. The areas of subsistence agriculture shown in Fig. 8.3 now represent areas where much of the rural food consumption is still produced directly by the consumer but where there is surplus production which is sold to feed the urban population. The money is then used to buy other items for rural family use.

Environmental associations in arid regions

Associations amongst distributions are usually used to define sets of regions. The regions may then be compared and contrasted, as in the example of agricultural types. Alternatively, the regions may be defined to allow more detailed study of one particular type of region. The study of individual ecosystems can be of this type. Arid regions, for example, have their own associations amongst climatic measures, plant and animal species, human occupancy, and land use types. The arid and desert areas of Africa illustrate this complex pattern of association amongst several factors. The usual distinction between arid areas and desert areas is made on the basis of rainfall totals, with desert areas having less than 100 mm and arid areas between 100 and 400 mm.

Within Africa there are three main desert/arid areas – the Sahara, the Somali-Chalbi desert and the desert areas of South Africa and Namibia. The Northern Sahara can be divided, on climatic grounds, into the area influenced by the maritime mediterranean climate and the interior area of continental climate. The influence of the Mediterranean is gradually lost towards the south, with the hot dry summers being replaced by extreme heat and no winter rainfall. The northern edge of the Sahara desert is usually considered as the 100 mm isohyet, or line of equal rainfall, but poor plant cover is still present, even with a rainfall as low as 50 mm. Generally the Saharan vegetation consists of relatively few species of drought resistant plants and small shrubs, with larger shrubs and trees in wadi beds where the water table is close to the surface. In the wadi beds a variety of plants may be grown, and agriculture is carried out with the aid of irrigation. The arid area of the Northern Saharan countries is very extensive, with well over 5 million km^2 of arid and desert lands in the five countries, as is seen from Table 8.2. The Southern Sahara is an area of tropical air mass influences (see Chapter 7), and here summer rains may only amount to 250 mm per

Table 8.2 Proportion of North Sahara countries in arid and desert zones

Country	Total area (1000 km^2)	Non-arid, non-desert (%)	Arid (%)	Desert (%)
Morocco	447	44.2	26.8	29.0
Algeria	2381	7.6	8.4	84.0
Tunisia	155	23.9	35.4	40.7
Libya	1760	0.2	5.1	94.7
Egypt	1000	0.0	3.0	97.0
Total	5743	7.3	8.7	84.0

year, resulting in a very extensive arid zone. The Somali-Chalbi desert area in Ethiopia and Somalia is more complex climatically, with strong topographic and coastal influences superimposed on the broad patterns of world climate. The vegetation is similar to that of the Southern Sahara with widely spaced herbs, grass and small shrubs comprising a sparse vegetation cover. Shrub and tree savanna occur where rainfall approaches 200 mm on the desert edge. At the southern fringe there is a rapid transition from desert to tropical forest.

The South African desert, comprises the Namib, Kalahari and Karroo deserts. The Namib desert has two distinct climatic districts. First there is a very dry, almost vegetation free coastal area, with less than 50 mm of rain; additional moisture from fogs and dews is caused by hot air off the land mixing with cold air at sea, over the cold Benguela current. Second, away from the coast the climate is less dry as the western boundary of the Kalahari is approached and short grasses appear after summer rains. The Kalahari desert has winter rains in the south and summer rains in the north where the desert margin has been extended by heavy overgrazing of the thin vegetation. Acacia dominates the vegetation both in tree and shrub forms. The Karoo desert is an area of winter rains and merges with the Kalahari desert.

Desert and arid areas in Africa also have distinctive land forms and soils compared with those of other ecosystems. Within the vast area of the Sahara four different kinds of landscape are recognised on the basis of land forms. These are:

1 Rocky and steep sloped hills or mountains
2 The *Reg* (also called *Serir* in the Eastern Sahara), which are gravelly-pebbley, almost horizontal plains
3 The *Hammada*, which again are almost flat areas but are covered in large flagstones of limestone, sandstone or basalt
4 The *Erg*, which are areas of sand dunes of various types including the crescent-shaped mobile *barkhans* and the huge fixed *ghowds*, which can be up to 200 m high and usually have a rock skeleton

The arid and desert regions are not defined and subdivided only in terms of climate, vegetation and land forms. They have associated with them characteristic land use and agricultural types. Table 8.3

shows the associations of these characteristics in the Sahara. The agricultural types are pastoral nomadism, irrigation agriculture and occasionally dry farming, which uses the wet season and residual soil moisture to grow crops.

Table 8.3 Vegetation and land use zones on the southern margin of the Sahara

Zone	Countries	Approx rainfall (mm)	Vegetation	Major land uses	Main crops/ livestock
Saharan desert	Mauritania, Mali, Niger, Chad, Sudan	less than 100	Desert	Nomadic grazing	Some camels
Sub-Saharan arid	Mauritania, Mali, Niger, Chad, Sudan	100–250	Subdesert steppe with grasses	Nomadic grazing	Camels, sheep, cattle
Sahelian arid and semiarid	Sengal, Mali, Mauritania, Upper Volta, Chad, Niger Sudan	250–600	Thorny shrubs, wooded steppe with grasses	Semi-nomadic grazing, limited dry farming and irrigation	Goats, cattle, sheep, sorghum, millet, irrigated rice

Agriculture in arid North Africa

Nomadic pastoralism is common in arid regions and is well-developed in the subSaharan and Sahelian zones (Table 8.3). There is great variety in detail amongst the different types of nomadic pastoralism, with some groups perpetually on the move whilst others move seasonally and may reside in one locality for several months each year. Rainfall is insufficient in the desert and arid zones to provide feed for domestic animals in one place without supplementing their food supply. Very few areas in North Africa have the necessary access to water to allow the growing of fodder, and so in the dry season livestick are ranged over large areas. At one extreme are the true pastoral nomads such as the Tibus of the Libyan desert, who live in small groups with only a few sheep and goats ranging over hundreds of square kilometres. At the other end of the scale, in areas where rainfall or moisture supply is higher, a permanent residence may be maintained for a few years, crops grown and a base provided for the animal herding activities. In some instances two bases are maintained, one for the dry season and one for the wetter period. The dry season base is located in an area of dry season grazing, whilst the wet season base may be in a place that allows limited crop cultivation. The regular seasonal movement of people and livestock in this way is called *transhumance*, and it is not only a feature of arid areas. The majority of nomads in North Africa, however, live between these extremes and have relatively fixed patterns of seasonal movement related to rainfall incidence. In the Sahel and on the southern margin of the Sahara, for example, the herds are at their furthest south during the dry season. At the start of the rainy season they move north following the rains and new grass growth. During the wet season they graze lands in the north and then return south to graze the main season growth of areas they used early in the wet season.

Dry farming in the North African arid zone is only possible either on the margins where rainfall is greater than 300 mm or in a few cases where ground water is available. Ground water availability more commonly results in irrigation agriculture. In dry farming terracing and mulching conserve the soil from erosion in the dry season and collect rainfall in the wet season. Typically, crops such as wheat or barley can be sown at the onset of the wetter period, but if the appearance of this season is not dependable, sowing may be delayed until the wet period has actually arrived. Growth can then be sustained by the slight but effective rainfall, and crops can be harvested before the dry season causes the ground to dry up altogether. The cultivated land in fallow must then be protected for the remainder of the dry season. There is a tendency for dry farming to extend into areas of marginal rainfall as pressures on the land increase through growing population numbers. Such extensions can be disastrous: pastures are destroyed, the dry farming fails and extensive soil erosion takes place as cultivation retreats again. The boundary area between dry farming and pastoralism is a very difficult area in which to apply conservation methods. As plant growth is very slow and species diversity is low, the vegetation has a low level of ability to recover from misuse.

Irrigation agriculture is a more stable and manageable form of crop cultivation in arid zones generally and in North Africa particularly. Two types of irrigation are commonly used. First is drawdown agriculture, when crops are planted after seasonal flooding. Second is irrigation proper, where river water is controlled or diverted to water a growing crop. The irrigation of arid and desert soils on flat alluvial surfaces usually involves an elaborate system of dams, canals and irrigation ditches. These have to be maintained, and decisions have to be made by the community as to how much water is allowed down the different channels and so to different areas of crops. The irrigated area is subject to very heavy water losses through evaporation and evapotranspiration. Salts contained in the irrigation water remain in the soil and gradually increase their concentration unless flushed out by using more water. The increase in salt in the soil is called *salinization* and in due course it creates a reduction in crop yields. A second problem is *waterlogging*, in which large volumes of water used in flat areas may cause saturation of the soil and again reduce yields. In this case drainage is necessary to remove the surplus water. It is possible, therefore, for an area of land to require both irrigation and drainage. Oasis agriculture depends on irrigation, and the natural Saharan oasis vegetation of tamarisk and oleander can be replaced by date palms, fruit trees, vegetables and fodder crops. The extent of agriculture in these oases is directly related to the yield of the water source and care must be taken to avoid overextraction. Cultivated land nevertheless constitutes only a small area of the North African arid zone.

Climate, vegetation, land forms, water sources, agriculture and society are all interrelated in arid and desert ecosystems. Changes in one, or worse, in several, of these factors can have disastrous effects on the balance and associations amongst the others. In recent decades in the Saharan and subSaharan areas, rapid population growth has increased the need for food supplies. The area under dry farming, producing particularly barley and wheat, has been increased by clearing new land in the mountains and in the desert margins. Soil erosion is an increasing problem: for example, Algeria loses 40 000 hectares of agricultural land every year from water erosion, and only slightly less from wind erosion. Overgrazing has also occurred, so that actual production is about twenty-five per cent of what it could be with better management methods. Woody shrubs have been removed for firewood so the capacity of the land to feed livestock has been reduced.

The associations amongst the various landscape factors in the arid and desert areas suggest some causal influences amongst the factors. Water sources influence agricultural types, for example, and so it is possible to look more closely at associations amongst factors to see how they interact together. In this way the geographer can begin to move from describing the landscape and associations within it, often with maps and simple statistics, to trying to explain why it is as it is. Later we will concentrate on explaining landscape patterns, and then we shall return to some of the associations looked at in this chapter and try to establish explanations for them and the resulting landscapes.

9

World Flows

Flows occur within the world distributions outlined in the last two chapters. At the broad world level the flows in general move energy, people, resources and goods from areas and regions of surplus to areas of deficit or, for economic goods, from areas of supply to areas of demand. Such a generalisation is only valid for the broad world pattern of flows, and when more local flow patterns are considered the picture is much more complex. The purpose of this chapter is to describe some of the major patterns of world flows. Three examples will be used: some aspects of ocean currents, world population migration and world trade.

Movement of ocean currents

The ocean currents are one of several means by which heat is moved from low latitude near the equator, where solar radiation is greatest, to high latitudes which are comparatively deficient in radiation receipt. The standard unit of measure of radiation from the sun is called a *langley*, which is equivalent to one gram calorie of heat received by one square centimetre of surface. Over the year this value varies daily according to the latitude of any part on the earth's surface. At the spring and autumn equinoxes the equator theoretically receives 890 langleys per day, whilst at the poles none is recorded. Alternatively, in January and December the pole having summer receives over 1000 langleys per day, whilst at the equator the value is almost 800. Thus polar variation is from zero to 1040, and at the equator the variation is only between 790 and 890.

Much of this radiation received at the outer edge of the atmosphere never reaches the surface of the earth, but is reflected from clouds and dust in the atmosphere; some is directly reflected back from the earth's surface and so cannot be redistributed within the ecosphere. The reflection loss for the earth as a whole is thirty-two per cent of incoming solar radiation and this is termed the *earth's albedo*. The remaining sixty-eight per cent of radiation is received at the earth's surface with, over the year, equatorial regions receiving about 140000 langleys and polar regions about 60000 langleys. Britain receives approximately 90000 langleys per year. The steady impact of radiation in equatorial regions is redistributed poleward, mainly by the ocean currents and the atmosphere. In the equatorial regions the oceans absorb higher levels of surface radiation than land areas because the amount reflected back is lower. The *albedo* of oceans is around six to ten per cent; for continental areas it ranges from five to thirty per cent, depending on the vegetation cover, and if the surface is snow or ice covered, albedo varies from forty-five to ninety per cent. The net result is that the areas of highest surface radiation absorption are equatorial oceans, particularly the mid-Pacific, and subtropical deserts.

The flows of the ocean currents effect a heat transfer between low and high latitudes. An ocean current is any persistent horizontal flow of ocean water and it is set in motion by prevailing surface winds. Winds move air from high to low pressure areas in the atmosphere, but due to the earth's rotation the flow of air is not direct from high to low pressure but is turned by a force called the *coriolis force*. The effect of this is that any object or fluid moving horizontally in the northern hemisphere tends to be deflected to the right of its path of motion. In the southern hemisphere the deflection is to the left. The coriolis force is absent at the equator and increases in strength toward the poles. Because of the coriolis force, water drift in a current is deflected either to the right in the northern hemisphere or to the left in the southern hemisphere. The major ocean currents are set in motion by the winds around the subtropical high pressure areas centred about 25° to 30° north and south of the equator (see Chapter 7).

Fig. 9.1 shows the generalised pattern of flows in a major ocean stretching from pole to pole. The patterns in the Pacific and Atlantic oceans follow this generalised picture. The main features of Fig. 9.1

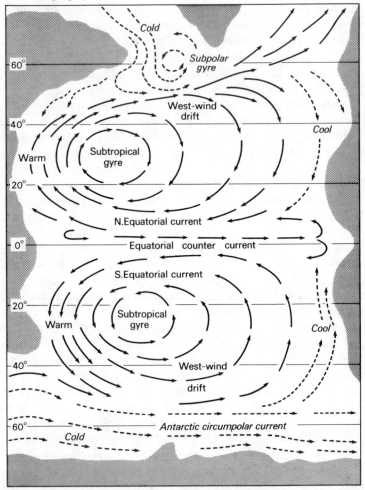

Fig. 9.1 The generalised flow pattern of ocean currents

are the pair of circular movements called the sub-tropical *gyres* which flow beneath the sub-tropical high pressure areas. In the northern gyre the flow is deflected to the right, and deflection is to the left in the southern one. In the equatorial regions the north and south currents move from east to west, are deflected and move warm water northwards and southwards respectively in the two hemispheres, past the east coasts of the continents. This movement of water towards the west of the ocean is offset at low latitudes by the equatorial counter current and at higher latitudes by the west wind drift. The warmer equatorial waters are transferred to higher latitudes, where they cool. A simple circular current around Antarctica mixes with the warmer equatorial waters brought south, the current cools and moves with the west wind drift to the eastern side of the ocean, where it acts as a cool current flowing northwards along the western continental margin. In the northern hemisphere the pattern is slightly more complex, for the simple circumpolar current is absent and the west wind drift splits and carries some warmer water to high northerly latitudes and brings some cooled currents southward along the west of the continental margin.

The general pattern of Fig. 9.1. is seen in both the Atlantic and the Pacific, where in the northern hemisphere warm currents are carried to the coasts of central America and Japan by the Gulf stream, and by the Japan current or Kuro Siwo. The North Atlantic and North Pacific drifts move the warmer water through the higher latitudes and being deflected to the right bring colder arctic water south past the North African Atlantic coast (Canaries current) and the western USA (Californian current). The North Pacific Drift also effectively brings warmer water to the Alaskan coast, whilst the North Atlantic Drift brings warmer waters to the British Isles and Scandinavia. The ocean currents, in turning around the great gyres, redistribute heat from areas of comparative radiation surplus to areas of radiation deficit.

Intercontinental population migration

The great intercontinental migration flows of the population have tended to be from relatively densely populated areas to relatively sparsely populated ones. This is not true of migration within major continental areas, where in historic times there has been a retreat from rural areas in favour of urban living. Major world movements of population in recent centuries have been from Europe and Africa to North America, and also from Europe to Latin America and Australia. There have also been major movements out of India, northwards from China, from China through South East Asia, and eastwards through Central Asia. The general pattern of these migrations is shown in Fig. 9.2.

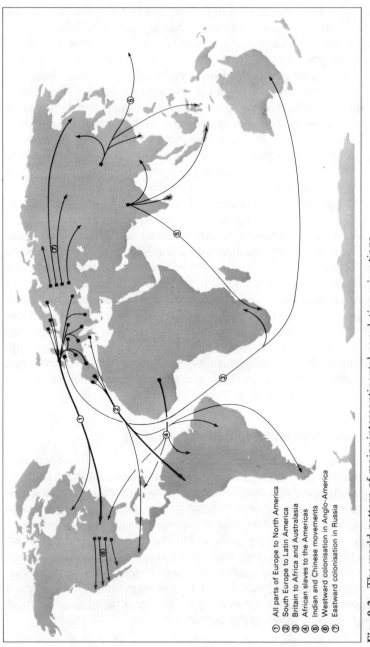

① All parts of Europe to North America
② South Europe to Latin America
③ Britain to Africa and Australasia
④ African slaves to the Americas
⑤ Indian and Chinese movements
⑥ Westward colonisation in Anglo-America
⑦ Eastward colonisation in Russia

Fig. 9.2 The world pattern of major intercontinental population migrations

These patterns of migration show the response of population both to the economic and social opportunities offered by frontier lands, sometimes termed 'pull' factors, and to conditions prevalent in the migrants' home countries, sometimes termed 'push' factors. The patterns of migration depend not only on the 'push and pull' factors but also on the technology of transport and the amount of information available about potential destinations. The mass movement to North America from Europe had to await the development of a shipping technology capable of moving the many migrants, and it also depended on potential migrants in Europe becoming aware of the opportunities available in North America.

Major migrations of population in recent centuries have been important in making possible the spread of settlement and the colonisation of new areas. Not all the movements have been voluntary, however, and there have been large scale movements of slaves and contract labour, for example from India to East Africa, for agriculture and mining. Such movements have encouraged, and sometimes been essential to, resource exploitation and economic development, and have had considerable subsequent economic and social consequences. Migration is also the principal mechanism for the spread of technology, language and general patterns of social behaviour. Australia, for example, would be quite different today if the main migration thrust had originated in China or India rather than in North West Europe. Migration can also take place in an attempt to escape undesirable social and economic conditions, the push factor, rather than as a positive search for new opportunities, the pull factor. There are several factors underlying the movements shown in Fig. 9.2, which is a very general map of migration patterns.

Several of the major migrations have originated in Europe, and among these the trans-Atlantic migration is by far the most important. Starting with small colonies of French, English, Spanish and Portuguese in the sixteenth century, the migration rate increased steadily until the early nineteenth century, when quite suddenly it became a mass movement of people from almost all the European countries. Until this mass migration probably only about two million people were involved over 200 years. From 1840 to 1914, however, more than fifty million people entered the USA alone. In the late nineteenth century European migrants were also looking toward other continents, with notable moves to South Africa,

Australia, New Zealand and South America. For several years after 1880, migration from Europe reached almost one million people per year. Although by far the largest, the exodus from Europe was not the only population migration of the late nineteenth century. Movement out from India, China and Japan also occurred as people sought better living standards in countries generally with a lower density of population. Because of the European control of both transport and the agencies of migration, even the movements out of Asia were often controlled by European companies.

Settlement resulting from migration

The migration of Indian labourers to the then Malaya (now peninsular Malaysia), West Indies and East Africa to work on plantations was managed by Europeans. The workers were indentured labourers who agreed to work for a fixed period for agreed wages in labour intensive agriculture such as sugar cane plantations. After their fixed period of hire expired many moved out of labouring jobs and opened shops or became traders, and sizeable minorities of Indian traders now flourish in these regions. The Chinese have similarly moved into trading activity after migration from their home country, and Chinese influence can be considerable in large cities in countries such as Malaysia and even in North America, where most large cities have a *Chinatown*. Around forty million Chinese now live outside China and much of the movement has taken place within the last 150 years.

The two main phases of European migrants are paralleled by two distinct periods of settlement in the new lands. In the initial phase up to the early nineteenth century, immigrants remained along the edges of the newly settled continents. The settlements tended to be of three types:

1　Trading stations, which served as collection and export points for luxury and exotic products of Asia and Africa. Numbers of settlers in these trading stations were small, but the settlements were important as points for the spread or diffusion of European culture.
2　Tropical and subtropical plantations, which encouraged production, in addition to trade, of luxury foods and industrial fibres

demanded in Europe. These settlements were often close to the coast but gradually were moved inland as transport facilities improved. Relatively small numbers of European managers were involved but often large numbers of immigrant labourers were employed either as slaves or indentured labour.

3 Farm-family colonies in middle latitudes became established, for example in North East America and later in Australia and New Zealand. These settlers tended to produce goods for local consumption rather than export to Europe and provided the base for the domestic economy of newly-settled lands. Again coastal locations were favoured initially.

The second phase of European expansion is characterised by a movement away from the coastal margins and a steady settlement of continental interiors. The settlements depended on the parallel growth of the European industrial economy and the growth of a domestic economy in the newly settled countries. Urban and industrial development in the settlements was necessary to provide for the rapidly growing population, which consistently moved inland with the increased exploitation of the mid-continental grassland for grain and stock production. The prairies of North America, pampas of South America, the Veld in South Africa and Murray-Darling Plains in Australia were all rapidly settled and turned into productive agricultural land. In each case European culture has been spread to other continents and in general the migrations have been from densely populated European countries to the sparsely peopled 'New World'. Broadly the movements have been from areas of labour surplus to areas of labour deficit. Even the more recent migrations to Britain from the West Indies, India and Pakistan can be considered until the mid 1970s and 1980s as movements from labour surplus countries to comparative labour deficit countries.

Types of population interaction

International population migration flows have had a major effect in redistributing world population over the last 150 years, but such flows represent only one type in a broad range of types of personal interactions. These types of interactions may be grouped into:

A   *Temporary* movements requiring no change of residence
B   *Transient* movements which require a temporary change of residence
C   *Permanent* movements.

In the first group it is possible to distinguish several major types of *temporary movement*:

1   Interactions between businesses
2   Interactions between consumers and businesses such as for shopping
3   The journey to work
4   Interactions for social purposes, such as journeys to school or for recreation or holiday purposes

Transient moves comprise, for example:

5   Military assignments
6   Movements of migrant labour which may be of any skill level as recent employment opportunities in the Middle East have shown, but more traditionally have been of agriculturalists (see Chapter 8).
7   College and university attendance, which usually affects a specific age group

Permanent moves can be of any length and not necessarily comprising one of the major international migrations discussed above. Major types are:

8   Migration for retirement
9   Migration for economic reasons such as to obtain a job
10   Migration for social and psychological reasons such as movement to live in the suburbs from city centres
11   Forced migrations usually for political reasons

Most of the world flows are of a transient or permanent nature and result in the creation of new landscapes which reflect both the adaption of migrant culture to the host environment and the environmental change introduced by new residents.

World trade flows

A third example of global flows illustrates other world patterns and features of the mechanism of movement. World trade takes place in almost all commodities and world patterns of trade differ with different commodities. Generally trade is greatest between countries with high levels of GDP per capita and a large industrial sector. World trade is sometimes divided into three types, with type one being trade between industrial countries. These are the major producing and consuming areas of manufactured goods, and consuming areas of the resources needed to support manufacturing. Europe and North America predominate in this type of trade.

Secondly there is trade between industrial and nonindustrial areas. The major industrial areas have their own traditional links with nonindustrial areas. This trade focuses on primary goods (food and industrial raw materials) moving from nonindustrial to industrial areas and manufactured goods moving in a reverse flow. Thirdly there is a relatively small amount of trade between one nonindustrial country and another. The world trade in oil fits this pattern less well than most other goods. The recent growth in importance of world trade flows in oil means that a third group of countries, the Oil Producing and Exporting Countries (OPEC), must be added to the industrial and nonindustrial or modern and traditional country dichotomy. This third group consists of countries that have relatively heavy trade, particularly in oil, with industrial countries, but often with only a slight reverse trade in manufactured goods because of the relative smallness of the markets in these countries.

The pattern of world trade both in volume and value is dominated by Western Europe. Many of the largest international flows of trade are between countries in Western Europe. Most of the other large flows have Western Europe either as a destination or a source. To some extent this pattern of a strong international web of trade in Europe is due to the contiguity of several countries – unlike, for example, North America where trade flows are internal rather than international. The pattern of flows can also be related to the idea of *comparative advantage*, which suggests that a country will specialise in the production and export of those commodities which it can produce at a comparatively low cost and import those goods which can be produced at a comparatively lower cost in other countries.

The idea assumes that there are no major barriers to trade between countries, such as an unwillingness for political or moral reasons, as for example, trade with South Africa. The principle is also concerned with trade in manufactured and agricultural goods, though it implies features about trade in the primary products needed for manufacturing. Trade is assumed to arise between countries due to differences in production costs. It is then possible, additionally, to consider transport costs in the trading operation, and other things being equal, more trade will take place between countries close together than countries far apart. The trade flows within Western Europe show this willingness of countries to trade with nearby countries, whilst the relative lack of trade across the border between East and West Germany illustrates the importance of political barriers to trade.

The idea of comparative advantage may be illustrated with a simple hypothetical example. Let us assume that car production in Japan can be carried out more cheaply than car production in Britain, but that jet engine production is cheaper in Britain. It would be to the benefit of both countries to concentrate on the item it can produce more cheaply – cars in Japan, jet engines in Britain – and to trade in the two commodities. It is not necessary that the cost advantages are absolute for the two countries, only that production costs differ comparatively and trade takes place. Suppose the following costs existed:

	Britain	*Japan*
Cars	130	90
Jet engines	110	100

In comparative terms, in Britain jet engines are cheap and cars are cheap in Japan. But despite an absolute advantage in both goods in Japan, it is still better for Japan to concentrate on cars and Britain on jet engines. With say 4 units of labour, material, etc., costs applied to the table for Britain and for Japan then minimum cost production in Japan is 360 cost units and in Britain is 440 cost units. This hypothetical example is obviously a great simplification of the real world pattern. It assumes free trade between countries, and also assumes both countries must produce something in order to create wealth. Inclusion of several countries and many more commodities makes a very complex pattern. This is made even more complicated by inclusion of the transport costs, trade barriers and the apparent ability of countries to run permanent deficits in their national accounts but yet remain in existence. The global pattern of world trade is described fairly easily but it is extremely difficult to explain.

The political aspect is always present in any consideration of international trade despite its essentially economic function of matching supply areas to demand regions. Given the economic rationale for trade, then political policies can be pursued, encouraging or discouraging trade with particular partners. Various trading blocs exist in the world which aim to encourage trade amongst themselves but discourage trade with those outside the bloc. The European Economic Community and the Latin American Free Trade Association are typical of many organisations with these aims. Internal barriers to trade, such as customs regulations and regulations or *quotas* on the amount of particular commodities which can be traded, are reduced, whilst a tariff wall is erected around the member states. The imposition of tariffs means that an outsider has to pay a levy to trade with a member country. Political activity such as occurs within trading blocs can be the single most important factor underlying the trade flows of a particular country.

Table 9.1 Value of goods in world trade – exports in million US dollars

	1938	*1948*	*1958*	*1968*	*1978*
Developed market economy	15100	36600	71400	168800	874355
Less developed market economy	6000	17200	24900	43600	156184
OPEC ·	1000	3100	7400	14000	145589
Centrally planned economy	1600	3700	12300	27300	125554
World	22700	57500	108600	239700	1301680
Per cent of trade in					
manufactured goods	46	52	59	64	63
fuels		10	8	9	17

(Source: United Nations Yearbooks)

World trade flows have increased steadily over the last century. Table 9.1 shows the enormity of the increase in recent decades. The great growth in world population since 1945 has been the main stimulus to trade expansion, but as can be seen from Table 9.1 the growth has been most marked in the developed countries which in general have had lower rates of population increase. The dominance of Western Europe and North America in world trade patterns has increased since 1945, such that in 1948 the developed countries accounted for sixty-three per cent of exports (excluding OPEC countries) but thirty years later it was over seventy-five per cent. Given the slight increase in the trading activity of the USSR, China and Eastern Europe, then most of the proportional loss in trade is accounted for by the less developed countries. This shift in contributions to world trade is paralleled by a change in composition with an increase in trade in manufactured goods and a decrease in the importance of primary resources (excluding oil) in world trade. Wool and rubber for example, which were major products in world trade before 1945, no longer feature as major world trade commodities but trade in synthetic fibres and petrochemical products has increased considerably. The massive growth in involvement of OPEC countries is also an outstanding feature of Table 9.1. These flows of fuels are associated with reverse flows of money, and one of the major problems facing the world economy is how to return into world circulation the dollars which are accumulating in OPEC countries.

As with the broad world economic and social patterns discussed in the two previous chapters, so also the major patterns in world economic and social flows are associated with the contrasts in levels of development in the world. At the broad world level the extremes of wealth which exist between Western Europe and the USA at one end and tropical Africa and South East Asia at the other result in differences in intensity of economic activity and fundamental influences on world patterns of interaction.

10

Patterns of Regions

The examples of world distribution and flows described in the last three chapters provide the backdrop for the description of landscapes. Often, however, when we look at landscape, we are more immediately aware of smaller scale features in the foreground. Whilst we are looking at the small units within global systems we are still interested in the distributions of features, associations amongst them and flows amongst them. On this smaller scale, however, geographers often try to identify generalised patterns against which to compare actual landscapes. We shall consider some of these patterns in this and the next chapter.

Urban land-use patterns

Land use and associated landscapes in a city have a systematic pattern to them. There is a general pattern of land use in cities, but the culture of the society and age of the city are important in creating the specific patterns of housing, offices, industry and open space. Urban land uses vary in their *function*, for example as between housing and its residential function and industry with its manufacturing function. Function defines what an area is used for. Land uses also vary as to the *intensity* with which the land is used for a particular purpose. For example, an area in the city may be completely residential or it may have in it patches of industry or open space or other uses. Intensity defines how concentrated a particular function is in an area. There is a third feature of urban land use and this is *form*. Form defines what structures comprise the land use of a particular function. For example, in an area of intensive residential function the buildings may be multistorey blocks of flats, as exist in many East European cities, or they may be individual houses each in their own plot of land, as in suburban America, or they may be squatter slums, as exist around many Third World cities. These three examples are present in quite different cultures – centrally-planned communist East Europe, affluent capitalist America and the poverty stricken Third World. The culture of the city is an important factor in understanding urban form. Similarly the age of the city is important. In British cities, for example, there are areas of intensive residential function built in the nineteenth century in terraces with relatively small rooms and to a very rigid layout. There are also large estates of housing developed in the 1930s with semi-detached buildings each in approximately 550 m² of land. More recently, since the 1960s, equally large estates have been built with much greater mixture of housing form, but with each house often having less than 350 m² of land. There are also many other age-related forms of the residential function in British cities. Other functions, such as industry or offices, have an equally wide range of forms associated with their different ages.

Generalisations of urban land-use patterns

If we consider city development within the West European and North American culture regions in the last 100 years, several general features of form, function and intensity can be picked out and specialised regions in the city can be isolated.

The three most notable attempts to generalise city land-use patterns are:

1 An attempt to relate land uses to *zones* around the city. This approach was initiated by an American sociologist, E. W. Burgess, working in Chicago in the early 1920s, but has been widely applied by geographers as a method of describing land uses.

2 An attempt to relate land uses to *sectors* with the city. This approach dates from the mid 1930s and is the work of H. Hoyt, again an American.

3 An attempt to relate land uses to several individual activity

points, termed *multiple nuclei*, in the city. This approach is based on work by C. D. Harris and E. L. Ullman, yet again Americans, in the early 1940s.

It is interesting that all three approaches were pioneered by Americans who were studying American cities. Attempts to apply the generalisations to European cities have met with less success. The patterns deriving from the three approaches are summarised in Fig. 10.1.

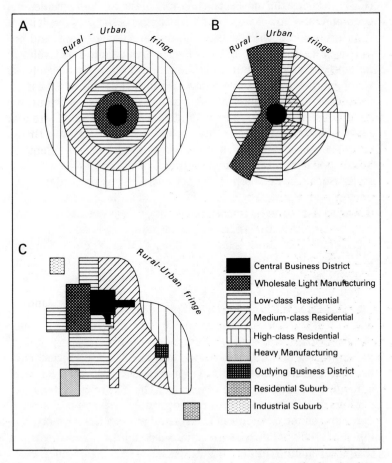

Fig. 10.1　Zonal, sectoral and multiple nuclear patterns of urban land use

The zonal pattern for the organisation of land use is also called the ecological approach, after the process which it is argued results in the zonal pattern. The process is one of invasion, succession, concentration, and decentralisation, with the development of the city taking place outwards, without land use planning, from the core. Urban functions move from the centre and invade new areas, become established and concentrated before being moved on and outwards (decentralised) as a result of the next invasion. As can be seen in Fig. 10.1 the zones begin with the central business district (CBD) which is the most accessible area of the city and the centre of economic and social life. The CBD is surrounded by an area of transition where older private houses are being taken over for offices or light industry or are being subdivided into smaller dwelling units. This transition zone is often an area of unstable social groups and rapid economic change. Around the transitional zone is a zone of working class family housing with more stable social conditions. A fourth zone of middle class housing exists even further from the city centre and beyond this is the commuter zone, which exists beyond the continuously built-up area of the city. Superimposed on this basic pattern are distinctive areas with particular social and economic characteristics. Thus particular ethnic immigrant groups, such as the Chinese and Italians, have their own districts; particular specialist functions such as residential hotels exist together in a district; and even particular residential forms such as apartment houses have their own district. These districts are present within the broad city zones. The generalisation of city zones is useful as a framework against which to match actual patterns of land use in the city. In all cities the patterns are more complicated than suggested by this generalised approach, but it provides a useful first step in trying to sort out and make sense of actual distributions of urban land use.

Dissatisfaction with the zonal approach led Hoyt to suggest that a more useful framework for description is one based on sectors of land use extending from the city centre. These distinctive wedges of land use, often based on a transport route, are extended at their outer end as the city grows. The general pattern is again shown in Fig. 10.1. The idealised pattern includes industry, which was not really considered in the zonal approach, but again the sector approach is most concerned with residential development. Within each sector the newest development is at the outer edge and so the age of buildings shows a series of concentric zones through a sector with the older buildings nearer the centre. The sector approach can then be seen as a refinement of the zonal pattern, with each sector containing its own sequence of concentric zones. In the sector approach the land use regions are associated with the transport network and it is assumed that this is focused on the city centre. Manufacturing is located along the sector containing the railway and the basis of the working class housing sector is a radial set of tramway or other public transport routes. When transport ceases to be focused on the city and cross-city routes develop then the sector pattern begins to break down and concentrations of different land uses emerge at key suburban intersections in the transport network. Fig. 10.2 shows this pattern schematically, with a restricted residential area, manufacturing along the railway and some suburban shopping districts developing in residential sectors at major tramway intersections.

Fig. 10.2 also shows the pattern after the appearance of mass car ownership. Tramways have been replaced by expressways and urban motorways, the area of residential suburbs has been extended and some of the key points of high accessibility in the suburbs have developed a mix of functions whilst others have become specialised areas. With the change in transport technology the zonal and sectoral patterns of land break down into a more dispersed pattern.

Both concentric and sector patterns are simple patterns which can be seen in land use maps of cities over the last 100 years. The multiple nuclei pattern of Harris and Ullman suggests a cellular structure to the city with distinctive land uses around certain dynamic nodes. This pattern is much more typical of present day cities with high levels of car ownership. Accessibility is not the only reason for cities developing this cellular pattern of land uses. In some cases there are distinct advantages in particular industries grouping together. These are called *economies of agglomeration* and they exist when firms or industries are linked together closely using articles produced by one another. Also some land uses shun each other, for example heavy industry and high class housing are seldom found close together.

The pattern of land-use concentrations in the multiple nuclei approach is due to particular land uses being preferred at particular

Pre-car   **Post-car**

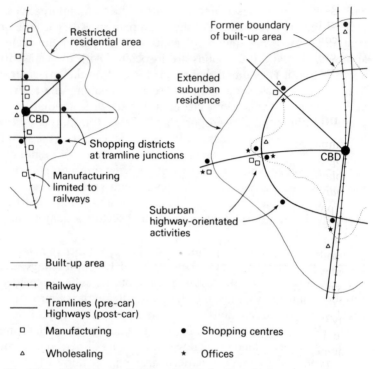

Fig. 10.2   The decentralisation of urban land use in the post-car age

types of location. Heavy industry is often attracted to a flat area of cheap land and high class residential development may be attracted to elevated areas providing a view and protection against hazards such as flood and urban pollution. The precise locations of the growth point nuclei will depend on the historical development of a particular city. The multiple nuclei scheme shown in Fig. 10.1 is a considerable simplification of real city land-use patterns. Heavy industry, for example, often operates at several small nuclei, some of which are currently growing while others, which grew several decades ago, are in decline. The multiple nuclei pattern, however, comprises a series of distinct land-use areas whose number and position depend upon the size of the city, its precise function, peculiarities of its site and historical development.

The patterns of land use in actual cities are more complex than suggested in any of the three generalised patterns outlined above. In order to understand why the land-use pattern is as it is, a necessary precondition is some form of simplification and the three types of generalised pattern provide yardsticks against which actual patterns can be simplified and compared.

Land capability regions

Comparison of patterns against a stylised scheme is commonplace in geography, both as an aid to describing patterns and ultimately in explaining them. In the last thirty years a number of methods of *land capability assessment* have been devised with the aim of describing the patterns of the potential, rather than actual use of land. These studies have been mainly of rural land, to assess the agricultural potential. The range of possible uses for a piece of land is limited by its soil, slope and climatic characteristics. At one extreme there is land which is so severely limited by its soil, slope and climate that it has no agricultural potential and can only be left as wildlife habitat. At the other extreme is land which has virtually no limitation and may be used for the intensive cultivation of a wide range of crops. The United States Department of Agriculture has developed a scheme of land classification in which eight *land capability classes* are distinguished. Class I has no limitations of any consequence, and Class VIII is wildlife habitat. The eight classes distinguished are:

Class I   Soils with few limitations that restrict their use
Class II   Soils with some limitations that reduce the choice of plants or require moderate conservation practices
Class III   Soils with severe limitations that reduce the choice of plants or require special conservation practices, or both
Class IV   Soils with very severe limitations that restrict the choice of plants, require very careful management, or both
Class V   Soils with little or no erosion hazard, but with other limitations impractical to remove, that limit their use

largely to pasture, range, woodland or wildlife food and cover. (In practice this class is used mainly for level valley floor lands that are swampy or subject to frequent flooding)

Class VI   Soils with very severe limitations that make them generally unsuited to cultivation and limit their use to pasture, range, woodland or wildlife

Class VII   Soils with very severe limitations that make them unsuited to cultivation and restrict their use to grazing, woodland or wildlife

Class VIII   Soils and land forms with limitations that preclude their use for commercial plant production and restrict it to recreation, wildlife, water supply or aesthetic purposes

There is a relationship between the level of limitation and the intensity of use. The limitations increase from Class I to Class VIII and the potential intensity of use decreases over the classes.

Each of these classes may be subdivided and the limitations specified. Limitations such as erosional hazard, soil depth, climate, stoniness, salinity or wetness are indicated by a letter. So, for example, land in Class V with limitations of excess water and a poor climate would be designated V*wc*. This scheme for describing the potential of land has been used widely in the USA and has also been used as a guide in less developed countries. Again, as with the urban example above, the scheme provides a framework for the description of landscape patterns.

Sequences of slope form

Regional patterns in land form description also make use of a generalised framework against which actual patterns can be matched. The description of slopes is the cornerstone to the identification of particular types of land forms. In the description of hillslopes, up to nine *slope segments* can be identified. In cross-section these segments are shown in Fig. 10.3. At the upper end of the *slope profile* are three convex elements, towards the lower slopes are two concave elements (segments 6, 7). These two groups are separated by two straight slope elements (4 and 5) and at the slope base are two more straight slopes (8 and 9) associated with a river and stream channel.

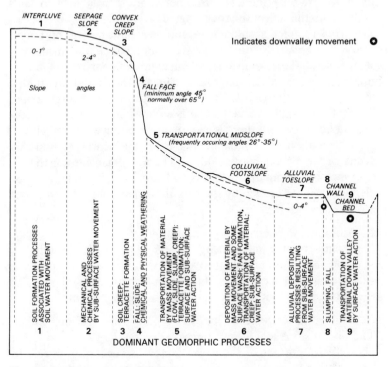

Fig. 10.3 Segments of a hillslope

The convex elements are shallow slopes of only a few degrees inclination. The three elements are usually termed *interfluve*, *seepage slope* and *convex creep slope*, with the names providing an indication of how rock and soil materials are moved on these elements. Soil creep is particularly important and accounts for much of the slow downhill movement of material. The rate of soil creep is highest at the surface and lower at below ground level. Leaning fence posts and trees with curved trunks are evidence of this process.

The straight segment of slope below the convex slope is called the *fall face* and may be very steep, appearing to be nearly vertical. Below the fall face there is usually a *talus*, or *transportational midslope*, containing coarse rock debris which is termed *scree*. The scree usually provides a shallow cover over the bedrock, and material in the scree moves continually down slope with replacement from material falling from higher slopes. The angle of the talus depends on the size of the rock fragments in the scree, the shape of these rock fragments, climate, vegetation and the rate of supply and removal of rock fragments to the scree. Large rock fragments form steeper scree slopes than smaller fragments and angular fragments produce steeper angles than rounded areas. The third straight slope in Fig. 10.3 is the *channel wall* at the base of the overall slope. Straight-sided gulleys and river cliffs occur where erosion is particularly rapid and where there is a lot of downcutting by rivers.

In the concave parts of the overall slope, flowing water is the main agent for the movement of material. The water may move in a sheet or it may be channelled into rills and streams. The concave slope elements will be eroded by the runoff of water and the amount of erosion depends on:

1 Amount and intensity of precipitation
2 Length and steepness of slope
3 Kind and amount of plant cover
4 Cultivation practices used if the slope is in agricultural use
5 Soil characteristics such as stoniness and clay content.

Generally the longer the slope the greater the erosion, with a doubling of slope length resulting in an increase of 1.5 in soil loss. The most important of these five factors is vegetation cover. In forested slopes, the rate of soil removal is often about the same as the rate of soil generation. With nonvegetated areas erosion rates can be very high, particularly with relatively unconsolidated soils and long steep slopes. On forested slopes erosion may be only five tonnes of material per sq km per year, whilst at the other extreme values may reach over 1000 tonnes per sq km per year. It is on these lower concave slopes that the drainage patterns discussed earlier in Chapter 2 occur.

The various slope elements can be mapped by plotting the pattern of breaks and changes in slope. A break in slope is a discontinuity in the slope profile; a change in slope is a much more gradual transition from one slope element to another. Breaks in slope are usually mapped as solid lines, while changes are shown by broken lines. Concave, convex and straight slopes are distinguished by marks on the downslope side of the line separating the two slope elements. It is also usual to measure the degree of slope, marking this on the slope map. An actual map of slope element distributions produced in this way can then be compared with the framework provided by the generalised sequence of nine slope elements. Using these methods it is possible to describe the distribution of slopes in an area and to pick out similar associations of slopes which may define a particular land form.

The creation of simple models

In the three examples taken in this chapter the attempted description of landscape has been based on the comparison of the actual landscape patterns with a generalised pattern. This approach is frequently used in geography as an aid to regional description and as a step towards explanation. The generalised pattern is often formed from the application of a hypothesis and theory which is thought to explain a particular landscape feature. Such generalisations are termed *models*. Thus in the case of urban land uses in the concentric zone scheme, the hypothesis underlying the approach is that land uses result from a process of invasion, succession, concentration and decentralisation and a model of expected land-use patterns is the result. In the generalised slope profile the hypothesis is that processes of geomorphological change differ at the various points in the profile. The various processes are very complex, but are summarised on Fig. 10.3 and a nine segment model of slope character is created.

The comparison of actual with the generalised pattern, or model, highlights features of the landscape which do not accord with the hypothesis or theory, and so the hypothesis can be changed and new features introduced to produce a better generalisation or sharper model of landscape patterns. Although this approach to landscape description can be used at any scale it is most commonly used at a medium scale rather than for the global descriptions discussed in the previous three chapters. This approach does not seek to explain the processes operating in the landscape but it does provide a better basis for the description of the results of these processes.

11

Patterns of Points

The comparison of actual patterns in the landscape against a generalised framework or model is a common approach to the geographical description of point distributions in the landscape as well as of regional patterns. The aim in both cases is to improve the accuracy of the description of patterns in the landscape. Only after accurate descriptions have been made can explanations be considered as to why the features are where they are. Accurate descriptions of point patterns may be based on the simple statistical and mapping methods discussed in earlier chapters or they may be based on comparison with a generalised, possibly theoretically derived pattern. The framework against which actual patterns are assessed may be a classification, a check list of factors, or some logically derived hypothesis. In each, however, a framework is provided against which real world patterns can be compared. This approach to description is valuable because it allows the geographer to consider some of the relationships between the distribution being studied and other landscape features. These interrelationships become very important as the geographer moves from description of the landscape towards explanation of the distributions of points, regions, flows and landscape itself. In this chapter some distributions of point features in the landscape are discussed and also illustration is made of how point patterns relate to patterns of flows and to patterns of regions.

Industrial location

In the location of industry the places of production may be considered as points and the pattern of manufacturing locations as a point distribution. Manufacturing processes require flows of raw materials, energy and capital to be concentrated at one site, together with a commuting flow of workers. The distribution of finished products from the manufacturing site similarly has a flow pattern. In considering industrial location, therefore, we have a point pattern of locations sustained by a complex, interacting, pattern of flows.

Raw materials and energy costs

Manufacturing industry requires raw materials and in many cases raw material availability is the major locational factor. In the iron and steel industry, for example, the location of manufacturing plant occurs at places where the raw materials (coal, limestone or dolomite and iron ore) can be brought together relatively cheaply (see Chapter 2). The costs of materials and of assembly then represent the major element in production costs in most manufacturing industries. Location at a place where material costs are low will give an industry a better chance of success than if located elsewhere. For the petroleum industries over three-quarters of manufacturing costs are taken up in raw material costs, and for food processing the proportion is approximately two-thirds. Even in industries where the volume of raw materials is relatively small, such as instrument or watch manufacture, about a third of costs is accounted for by raw materials, which still represent the largest single item in the cost structure.

Raw material costs are of two types. First there is the cost of acquiring the materials, and secondly there is the cost of transport. The first of these costs only affects the location of the processing plant when there are significant price variations in supplies from different sources. Increasingly these price differentials are being reduced or eliminated. Many raw material prices are fixed, irrespective of source, by groupings and controls in the supplying industries. The Oil Producing and Exporting Countries (OPEC), for example, operate in this fashion and oil prices are fixed at regular meetings of the group, although political pressures in the group sometimes result in a breakdown in this unified pricing policy. Another reason for the source price of raw materials becoming less important as a location factor is the policy of many large companies to secure their supply sources for raw materials and components by extending ownership to include the major sources of supply. Such a pattern can be the case in the motor vehicle industry, where the sources of components have become integrated within a single motor vehicle producing organisation.

The second element in raw material costs is transport. This is a much more important consideration and can be critical in decisions on both industrial location and choice of raw material supply source. Transport costs vary in general according to distance, but there are other factors such as volume shipped and transport technology which influence cost per unit distance. Transport costs are of sufficient importance to warrant consideration as a factor of production and distribution in their own right (see below).

The major share of fossil fuel energy in an economy is consumed by manufacturing industry and service sector activities. In the British economy only a quarter of the energy produced is consumed domestically, the rest is used by industry and services. When energy was relatively immobile, for example water power from streams and rivers, industry was confined to sites where power was made. But as energy and energy fuels have become more mobile through electricity transmission lines and gas and oil pipelines, so power sources have played a steadily decreasing role in industrial location decisions. Transmission of electricity through a national grid makes this energy source effectively ubiquitous throughout a country. In a few cases involving industries requiring very large amounts of energy there are cost advantages to be gained by locating at the power source and saving transmission costs. Aluminium and copper processing and the production of fertilisers are industries of this type, and the concentration of aluminium production in the Pacific North-West and in the Tennessee Valley of the USA is a response to slight energy cost advantages, usually due to the availability of hydroelectric power.

Energy is also consumed in factories as well as in the production processes. Factories have to be heated and lit, and in this respect there can be substantial differences in energy costs from place to place depending on climate. In the USA, industrial costs in California and Texas are less than in Illinois and other states in the North East industrial area, but in some cases the lower heating costs are offset by the need for summer air conditioning. For the most part these variations are reflected in operating costs rather than acting as constraints or stimuli on the initial location of an industry.

Capital costs

The second major factor underlying the location of industries is capital availability and cost. Capital includes both the fixed capital of physical plant and financial capital. Fixed capital assets such as factory buildings are relatively immobile and are an investment at a site which often has a series of owners who perpetuate an industrial location pattern. This immobility leads to *inertial locations* of industry, with new firms occupying old premises. Often this type of industry changes with reoccupation, particularly if the original industry is in decline and so new industries become located at sub-optimal locations because of the presence of low cost premises. One of the best examples is the former cotton manufacturing region of Lancashire, where mill buildings have been taken over by other industries. Much the same happened in Japan in the 1950s when the factories of the declining silk industry were taken over by camera manufacturing and other similar industries.

Financial capital is quite mobile, although the costs of money through interest rates vary considerably from country to country. Even within one country the history of the development of manufacturing shows that local bankers and investors have played an important role in establishing industries in particular places. In the industrial development of New England money made by local shipping organisations and invested in local banks played a considerable part in financing the early industry. In the present day economy in general, large companies have little difficulty in obtaining finance for development but this is not always the case with small firms. Industrial groups in which small firms are widespread may have relatively low rates of investment and are particularly prone to the whims of local bank branch managers.

Labour costs

Labour is a third important factor influencing the location of
economic activity. The amount and type of labour required by
activities varies considerably over the different branches of the
economy. Many service sector industries have a very high labour
input, with wage costs representing the major component in overall
operating costs in most of the tertiary sector. But labour requirements alter, with technological change affecting both the total
amount of labour required and also its necessary skills.

Particular industries require distinctive types of labour, so making some locations more desirable than others. If workers of a
specific skill for an industry are required then there may be advantages in a location within an existing concentration of that industry.
In this way less training would be required but the competition for
labour could be increased, perhaps giving rise to higher wage rates
and greater labour costs, but labour productivity would be higher.
As in most industrial location decisions a delicate balance has to be
achieved, in this case between costs of training and wages. An
industry requiring a large labour force may find this easier to obtain
in a large metropolitan area or conurbation. Since labour is relatively mobile it is possible for a large employer to attract workers
from some distance by providing better working conditions, fringe
benefits and sometimes higher wage rates than other industries
competing for the same labour. Improved working conditions,
recreational facilities, welfare benefits, company subsidised transport – all add to overall labour costs. As metropolitan workers
would probably have greater expectations of these benefits than
workers in smaller towns, so total labour costs in large cities can be
higher than at other locations.

Labour cost is difficult to evaluate as an element in total industry
costs because it is complicated by the idea of productivity. Actual
labour costs may be high, but if there is a low absentee and low
labour turnover rate and high levels of productivity, then real costs
may be lower than in other industries with lower apparent costs but
lower productivity. Manufacturers in country towns away from the
big cities in Australia claim, for example, that absentee rates and
turnover rates are much lower than in large cities and so the high
costs of training are offset. But, in this case, these attractions are not

rewarded by higher wage rates in the small towns. Labour costs are certainly an important element affecting industrial and commercial location but their role is difficult to evaluate. There is a tendency, through changing technology, to substitute capital for labour and in the process change the structure of the employed labour force. Labour costs may well be of reduced importance in future locational decisions, but there is the possibility that the availability of workers with special technological skills may become more important. The residential preference patterns of these key workers could play an increasing part in influencing industrial location. The growth of technology intensive industry in Florida, Texas and California may be partly a response to a desire of workers to reside in the sunbelt and as such is an illustration of the power of labour to attract industry.

Transport costs

Transport is a further factor underpinning the location of economic activities. Transport costs pervade all analyses of activity location and are the costs of the flow network in industrial location patterns. These costs affect what a manufacturing firm pays for its raw materials and the price for the goods paid by the final consumer. The transport costs vary considerably with type of industry from around thirty per cent in pig iron manufacture to only a few percentage points in low bulk industries such as precision engineering or jewellery production. There are two aspects to transport costs as a locational factor. First is the type of transport used, and second is the freight rate structure. Locational decisions are made taking into account the balance between these two elements. A location involving longer distance transport, for example, may be feasible if it can make use of a particularly low cost transport *method* or *mode*.

Decisions on transport mode are influenced by prevailing technology. As road transport technology has changed, both with the development of more efficient and specialised trucks and a better surfaced and interconnected road system, so greater use has been made of this mode by all sectors of the economy. In the USA, Europe and Australia, for example, most country towns once had local bakeries and local breweries supplying the local market. Transport technology in the 1930s was such as to make local production of these items almost essential. The raw materials could be brought in but the finished product could only be moved a short distance without deterioration. Changes in transport as well as manufacturing technology now allow large scale production at a few sites and distribution by road over a wide area. Breweries in small towns have been closed and beer manufacture has been concentrated at a few sites with distribution, by road, to the national or regional market. The distance over which road transport can be efficient has increased.

The changing importance of a transport mode with evolving technology is seen with bulk transport. The larger size of oil tanker and consequential lower unit transport cost of crude oil is one of several factors which have been responsible for the trend towards market, rather than oilfield, located oil refineries. Large scale bulk transport of iron ore has allowed the exploitation of relatively remote ore sources by considerably reducing transport costs per tonne of ore. The opening up of major iron ore mining projects in Liberia is partly due to their relative proximity, by large bulk carrier, to major markets in Western Europe and North America. Western Australian shipments of ore to Japan are similarly dependent on large bulk carriers.

Generally the cheapest means of transportation for long hauls is water borne. Railways are cheaper over medium distances, and road transport cheapest for short distances. This general relationship is due to the relative balance between *terminal* and *line-haul* costs. For road transport, terminal costs are low but movement costs per unit of distance are high, whilst for water transport the reverse is the case. The detail superimposed on this broad generalisation makes the relationship between mode and cost much more complex in reality when applied to a particular product. For example, in the movement of oil in Western Europe, pipeline transport competes with rail and a variety of water borne modes depending on the size of pipe: larger diameter pipelines are competitive with ocean tankers over long distances. The decrease in cost for increasing distance is largest in the case of coastal tankers, which puts this mode in competition with road, rail and pipeline depending on the distance transported.

Within any one mode of transport, different types of *freight rates* may be charged. The discussion above relates to costs, but transport operators may recover these costs in a variety of ways through different pricing schemes. There are three main kinds of freight rate structure – *postage stamp rates*, *blanket rates*, and *kilometre rates*. The first involves the application of a standard charge irrespective of distance, as with mail deliveries. This type of freight rate has little effect on the location of activities. The second method involves the application of different transport rates to different zones. Zones can differ in size. More distant zones, if for example they allow large load delivery at a major market, may have lower charges than nearby zones, although usually more distant zones will carry higher charges. Under the kilometre rate system a charge is levied per kilometre, so that long hauls cost more than short hauls. The charge may be a constant or may decrease with distance. Both blanket rate and kilometre rate schemes effect locational decisions when minimum cost sites are sought.

The influence of the market on industrial location

The traditional factors of raw materials, capital, labour and transport provide the basis for understanding the location of economic activity. But goods are produced to be consumed, so the presence, size and location of the market for the goods can influence where the goods are produced. It is widely suggested that in many industries the significance of the market is increasing in relation to other factors such as labour, energy and raw materials. Cost differentials have decreased for some factors, such as labour. Other factors show increased mobility and less locational effect, for example energy and capital. With these changes the influence of the market increases. One of the main reasons for rapid industrial growth in North-East USA from 1945 to 1965, and currently in the densely populated area of Europe, is the ready market that exists for manufactured goods. This market is not just the final market represented by the population concentrations in these areas, but is also the industrial market of other manufacturing companies both in the producing country and in nearby countries. The high levels of international trade flows in Europe have been pointed out in the previous chapter. The increasing attraction of metropolitan areas is not limited to Europe and North America. Within tropical Africa the greater part of recent industrial growth is based on production for local demand and often involves manufacture from imported raw materials. There are some advantages to be gained from locations close to raw material sites or the points of import, but usually the chosen location is one from which the national market can be supplied. This is usually the capital city, or its immediate region, where there is the largest concentration of purchasing power. Combined with the market consideration in this case, there are probably advantageous transport links to demand nodes elsewhere in the country and also a large pool of labour. The largest industrial centre in Nigeria is Lagos, and in the Ivory Coast it is Abidjan, both large centres. Development close to, but not in, the major city is typified by the concentrations of industrial growth at Thika (35 km from Nairobi), Kafue (40 km from Lusaka) and Tema (30 km from Accra).

New technologies can create new markets for known products, with the resulting upsurge in demand stimulating production. Successive changes in iron and steel making technology over the last 200 years have opened up markets to a variety of types of iron ore. The Bessemer process required phosphorus-free ores and until 1879 there was a limited market for phosphatic ore. The Gilchrist-Thomas process changed the market demand considerably and allowed known deposits of phosphatic ore to be mined. More recent technological changes such as beneficiation, which increases the effective iron content of the ore by preprocessing the raw ore, have allowed low grade ores to be used successfully, again stimulating or prolonging ore production at particular locations. The low grade ore in northern Minnesota continues to be mined partly due to this technological change sustaining the market for these ores.

Governmental influences

Finally, it is necessary to consider government as a factor influencing the location of industrial activity. State and local taxes represent an element in production costs. This factor appears to be quite important in the analysis of alternative locations within a metropolitan area, but of less consequence in interurban analyses. In the intrametropolitan situation, where often many of the other elements in production costs are very similar in alternative locations, the difference in potential tax payments on inner and outer urban sites might be the critical variable tipping the balance of the locational decision, usually in favour of the suburbs. There is evidence from both New York and Philadelphia that tax levels for industry are substantially lower in the outer suburbs than in the inner city.

On a broader scale, attempts can be made to attract industrial activity to the less prosperous regions of a country by a system of tax incentives, cash grants, subsidies, and so on initiated by government. Most European governments have put such policies into effect in attempts to stimulate industrial activity in areas of high unemployment. The Development Area policies in Britain seek to encourage new industry into the older industrial areas such as South Wales, Merseyside, Clydeside and North-East England. Additionally special incentives are available for industries moving to, or setting up in, the remoter rural areas and in depressed inner city zones. Similar types of policy operate throughout the EEC, improving the locational attractions for industry in North Jutland, Southern Italy or Western Ireland and the central parts of major European cities. Government is an extremely potent force in creating and controlling large scale shifts in the spatial structure of manufacturing industry.

At the local level government has almost total control, at least in Britain, over the location of manufacturing activity through land-use planning policies (see also Chapter 16). Zoning regulations elsewhere in Europe and in North America are less stringent, but certainly act as a negative control over many locations. Such policies effectively forbid the location of some activities at potentially profitable sites. But in Britain both the creation of industrial estates and the refusal of planning permission at locations where development would run counter to the authorised land use plan provide a carrot and a stick to many of the decision makers involved in industrial location. The strength of government in location activity may over-ride all other factors. In much economic activity the administration of government policy, not just in Britain but in all free enterprise economies, provides the key to understanding locational patterns and spatial structures.

The description of the locational pattern of a particular industry may be related to these six broad factors. In some patterns raw material influences may be uppermost in providing a framework for the locational pattern. In other industries the market may be paramount. But for many industries all the factors will play some part, major or minor, if an accurate description of industrial location is undertaken and the point pattern of locations must be described both in terms of the distribution of points and of the pattern of flows to and from these points.

Central place theory

The general framework of factors underpinning industrial location provides the basis for the description of industrial patterns, but such a framework is extremely generalised. In contrast to this general approach there are more specific approaches to description which use logically derived frameworks against which reality can be compared. One such framework relates to the assumed spacing and location of settlements and is called *central place theory*. Working in the early 1930s a German geographer, Walter Christaller, suggested that there was a basic order and pattern in the location of towns. Towns differ from one another in respect of the degree of specialisation in the services they offer. Some towns will have only general services, other towns may have very specialised services. In the case of health services the *low level*, general service, towns might have just one doctor and surgery; other towns at a *higher level* will have more specialised services, such as a health clinic; and a few towns will have the complete range of general and specialist services, such as a full hospital. Not only health services are provided by each town, but also a range of retail and administrative services. Towns may be grouped according to the mix of services they provide, with towns of a higher level providing services for towns of a lower level, plus some extra services which are associated only with the towns in the higher level. The towns provide the services not only to the residents of the town but also to people living in the surrounding countryside.

Country dwellers travel to the nearest town that provides the service they require. The area around a town from which consumers of services travel is called the town's *complementary region* (or *tributary area*) and the attractiveness of a town due to the combination of services provided is termed the *centrality* of the town. Towns with a high centrality have more specialist services compared with the town of low centrality. High centrality towns have larger complementary regions and are more widely spaced than low centrality towns. Christaller's work in Southern Germany showed, as is seen in Table 11.1, that whilst market hamlets were on average 7 km apart, regional capital cities were 186 km apart. There were seven types of settlement with each type comprising places with a similar level of centrality. Centres at successively higher levels were thought to serve three times the area and three times the population, which made them $\sqrt{3}$ times further apart. So in Table 11.1 the complementary regions of successive levels are approximately three times larger, and the distance apart of the second level places is 12 km ($\sqrt{3} \times 7$). An American geographer, John Brush, carried out similar work in Wisconsin and found that hamlets there occurred on average 6.8 km apart, villages 15.8 km apart and towns 33.9 km apart. Many other studies, particularly in rural areas where the primary function of settlements is to provide services, have shown a similar regular spacing of places.

Table 11.1 The groups of settlement found by Christaller in South West Germany

Settlement form	Distance apart (km)	Population of place	Complementary region size (km^2)	Population of region
Market hamlet (*Markort*)	7	800	45	2700
Township centre (*Amtsort*)	12	1500	135	8100
County seat (*Kreistadt*)	21	3500	400	24000
District city (*Bezirksstadt*)	36	9000	1200	75000
Small state capital (*Gaustadt*)	62	27000	3600	225000
Provincial head capital (*Provinzhaupstadt*)	108	90000	10800	675000
Regional capital city (*Landeshaupstadt*)	186	300000	32400	2025000

The system of levels of settlement is called a *settlement hierarchy* and the hierarchies which increase by a factor of 3 are called $k=3$ hierarchies. Given the factor of 3 and the need for complementary regions for settlements at any one level to be the same size, then a logically defined pattern of places and complementary regions results. This pattern of points and regions is shown in Fig. 11.1. The diagram shows, for simplicity, only four levels of the hierarchy. The city has the largest complementary region, and at the edge of this region are six towns. Each of these towns has its own complementary region for the services it provides. The main city also has a complementary region for the town level services it provides because it is a town as well as a city. The parts of the town level complementary regions which fall in the city complementary region add up $[6 \times \frac{1}{3} + 1]$ to 3. This is the $k=3$ hierarchy. Each town is surrounded by six villages and again the rule of 3 applies to the areas of the complementary regions. This is the case for each level in the $k=3$ settlement hierarchy.

Other hierarchies are possible and Christaller suggested that where the cost of transport is very important a $k=4$ network results. Where administrative control is decisive a $k=7$ network is likely. The variation in k is critical in determining the spacing between centres and the number of centres at each level in the hierarchy. The lower part of Fig. 11.1 shows the pattern for just two levels of centre on the $k=4$ and $k=7$ schemes. In the three schemes $k=3$, $k=4$, $k=7$, the number of settlements in each of the first four levels of the hierarchy assuming level 1 to be of highest centrality are:

Level of Hierarchy	$k=3$	$k=4$	$k=7$
1	1	1	1
2	2	3	6
3	6	12	42
4	18	48	294

Central place theory has been extended from these initial ideas of Christaller to more complex formulations which allow more than one k scheme to be present in a single region and for other k networks ($k=9$, $k=12$, $k=13$) to be explored as frameworks against which to compare actual patterns of settlement size and location. Within central place theory a pattern of points and associated pattern of regions is generated from some theoretical assumptions

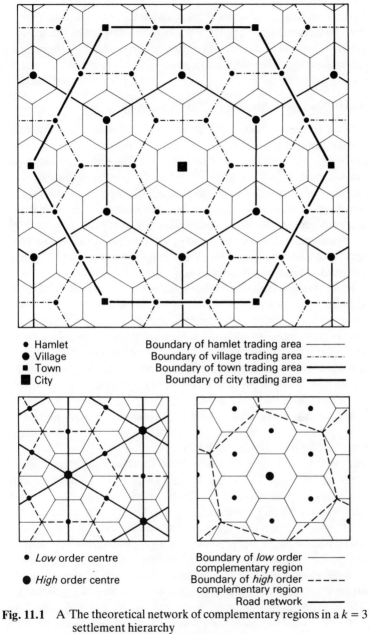

Fig. 11.1 A The theoretical network of complementary regions in a $k = 3$ settlement hierarchy

B Settlement patterns under $k=4$ and $k=7$ schemes

about how shops and services are provided and are used by consumers. To understand the pattern of points, in this case settlements, it is also necessary to understand the pattern of functional regions associated with the places.

Patterns and types of volcanic activity

The comparison of actual patterns against a classification of features rather than a theory, as in central place theory, may be illustrated by the study of volcanic land forms. To describe volcanic landscapes, however, it is again necessary to see the relationship between patterns of points, regions and flows. Volcanoes are characterised by eruptive activity when hot gases, liquids, molten rock and shattered earth fragments are forcibly ejected from openings in the earth's surface. Lava is the name given to the molten rock and most lavas are *molten silicates* with the range of silica (SiO_2) content being usually somewhere between thirty-five and seventy-five per cent. If the percentage is high, then the lava is termed *acidic* and is viscous; if the percentage is low the lava is termed *basic* and it flows freely. This distinction is important in determining how land forms are created, because basic lavas produce lower features covering a wider area than acidic lavas. Basic lava flows have a temperature usually between 1100 and 1200°C, and below this temperature they solidify. The upper surface of the lava flow is usually very *vesicular* or porous, because of gas bubbles escaping during cooling, but other surface textures also occur. If cooling produces an angular blocky surface the term *aa* is used to describe it, and if the surface is smoothly twisted like rope then it is termed *pahoehoe*. Both these terms come from Hawaii, where basic lava flows are extensive both from surface and from underwater eruptions. With underwater eruptions *pillow lava* results as blobs of lava cool on the outside in contact with the water but have liquid interiors.

Table 11.2 shows the classification of types of volcanic land forms. The type of material erupted varies from a fluid basic composition to extremely viscous lava with a large proportion of gas. The extrusions of basic material result in lava flows, or if the amount of material erupted is large, in basalt plateaux covering many hundred square miles. With relatively small eruptions low volcanic domes, often termed *shield volcanoes*, are formed. These are typically 100 to

Table 11.2 Types of volcanic activity and associated land forms

Type of lava	Type of Activity	Quantity of Eruptive Material			
		Small		*Great*	
Fluid, very hot, basic	Effusive	Lava flows	Domes	Basalt plateaux and shield volcanoes	
(Increasing viscosity, gas content and silica percentage)	Mixed	Cones with flows	Composite cones or Strato-volcanoes		Volcanic fields with multiple cones
		Domes with *lavaflows* (plug domes, spines, etc)			
Viscous, relatively cool, acidic	Explosive	Maars of thephra		Collapse and explosion calderas	
Extremely viscous, abundant crystals	Explosive, mostly gas	Gas maars	Explosion craters		

500 m high, but with a base diameter about twenty times larger than their height. Such shield volcanoes typify much of the volcanic activity on Iceland. The shield volcanoes of Hawaii are much larger. The two major ones, Mauna Kea and Mauna Loa, are over 4000 m high, and if the submerged portion of the volcanic group is included then overall the feature has a height of over 10 km. Larger still are the vast basalt plateaux where eruptions of lava have occurred through long fissures rather than from central vents. Most of these massive features are very old. The largest in North America is the Columbia Plateau, which is 130000 km^2 in extent. In India the Deccan Plateau is now 260000 km^2 and originally may have been twice that size. The basalts of the Deccan lava flow are over 2 km thick in places. Except for occasional small cones, the basalt plateaux are almost flat due to the highly fluid nature of the very basic lava of which they are formed. Whole landscapes evolve on large plateaux of this type.

With slightly more acidic material volcanic cones are formed either individually or in volcanic fields with multiple cones if large amounts of material are erupted. Ash and lava eruptions are typical of this mid-range of silica content material. In composite volcanic cones layers of lava are mixed with layers of ash which bury previous

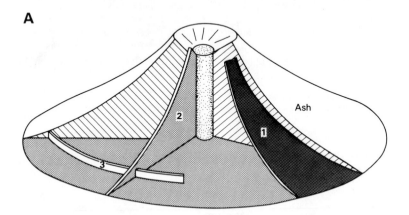

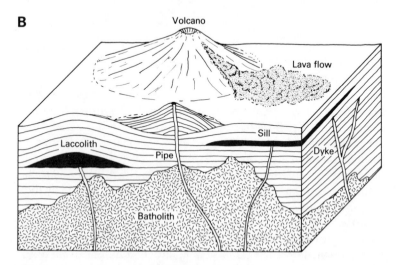

Fig. 11.2 A The structure of composite volcano
B Forms of intrusive volcanic activity

lava flows. Fig. 11.2 shows a generalised composite cone with a lava flow buried by ash and also eruptions of lava type material from long cracks around the volcanic cone. A composite cone usually consists of several alternative layers of lava and ash and lava erupted from cracks. These cracks may be circular around the central vent, resulting in *ring dykes*, or radial to the cone, so producing *radial dykes*. The ash buried lava flows are called *sills*. Dykes and sills are often conspicuous landscape features as they are left after the ash has been eroded. The cooled lava in the dykes and sills may be much harder than surrounding rocks and in time stand out as hills and ridges. The volcanic cones have straight steep sides and this type of feature is typical of much of the volcanic activity around the Pacific Ocean, for example in Indonesia, Japan, and Andean America.

With the more acidic lavas, *plug domes* and *maars* are formed. With plug domes the viscous lava is too stiff to flow and builds up as a dome or spine as it is erupted from a central vent. The spine that extended out of Mount Pelee in the Caribbean after its famous eruption in 1902 was of this type. *Maars* are caused by even more acidic lava which erupts by explosion and blasts out a broad shallow crater below the former ground level. Very large maars are called explosion craters or *calderas* and may be 10 km or more across. Crater Lake in Oregon lies in a caldera 10 km in diameter and 1100 m deep caused by a massive explosion 6700 years ago – relatively recently in geological time. Even more recently, the explosion of the volcano at Krakatoa in Indonesia in 1883 was of this type and a caldera 6 km in diameter was left as 80 km³ of material erupted from the volcano. A smaller explosion of the volcano Katmai in Alaska in 1912 was heard 120 km away and a caldera 5 km wide and 1000 m deep was produced.

From Table 11.2 a classification of all these volcanic land forms is possible related to the acidity of the lava and the quantity of erupted material. Volcanic land forms are the result of the extrusion of rock from below the earth's surface. Some land forms are created by the intrusions of hot rock which never reach the earth's surface, except perhaps after many million years when the surface is eroded to reveal the former intrusions. Both sills and dykes may be underground features which are later exposed by erosion. The lower part of Fig. 11.2 shows some of these intrusive features and a composite volcano.

Intrusive volcanic land forms

Batholiths are the largest of the intrusive forms and may be as large as 40 000 km² in area when exposed. The Idaho batholith in the USA is of this size and Dartmoor in South West England is a rather smaller, now exposed, batholith. Batholiths are often made of granite, but the smaller *laccoliths* are more likely to be made of basalt similar to extruded lava flows. From Fig. 11.2 it can be seen that laccoliths are close relations to sills, but with laccoliths the pressures created by the intrusion causes a doming of the overlying rock. Again as the surface is removed through erosion the laccolith may become exposed.

Volcanic activity is closely associated with the edges of the massive plates which make up the earth's surface. The discussion and figure of plate tectonics in Chapter 3 shows this association. The classification of volcanic activity shown in Table 16.2 provides the framework for the description of volcanic land forms. Although there are only about 550 known active volcanoes on the continents they produce a considerable variety of active land forms and sometimes rapidly changing landscapes. It is possible to describe both present day active volcanic activity and older extinct volcanic forms by reference to the classification in Table 11.2. The pattern of volcano vents may be mapped as a point distribution and, on the world scale, shows the association with the margins of tectonic plates. The landscapes associated with this point pattern have to be described by flow processes and formal regions in which the land forms result from volcanic activity.

The descriptive framework of volcanic land forms results from a classification involving type of lava, type of activity and quantity of eruptive material. The descriptive framework provided by central place theory is a logically derived pattern of settlement location and size. Actual landscape patterns may be compared against the forecasts of what should, according to classifications or theories, be present. The descriptive frameworks discussed in this chapter are all models. Models are generalisations of landscape patterns and processes which may be used as aids to description or as methods for analysing processes. The models in this chapter are mainly useful in enhancing the description of landscape. Many attempts have been made to model various aspects of the landscape. Models exist of agricultural as well as industrial location, of atmospheric processes, weathering and erosional patterns and processes and of many other landscape features. As has been stressed throughout this section of the book, landscape description is an important part of geography, but the ultimate aim of the geographer is explanation. Much of the explanation of landscape patterns and processes undertaken by the geographer depends on making models of how we think things have come about and then describing and surveying the landscape feature being studied to see how well our model fits reality. If it fits then a step toward explanation has been achieved. If it doesn't fit we must ask 'Why?', improve the model and then test it again. The next section of this book is concerned with explanation. The idea of using models is implicit to provide both the preliminary descriptive exercise and the analysis needed to yield an explanation as to why the landscape around us looks as it does.

PART FOUR
Explaining the Landscape

12

Cycles

In order to explain the patterns and variety in the landscape it is necessary to know something about the processes which create landscape change. Models of these processes can be thought out, tested, and if found to be valid used as the basis for explaining changes in urban and rural landscapes. The ideas of systems discussed earlier in this book help considerably with the building of these models of processes. The ideas of system inputs linked to outputs by internal processes is a first step towards analysing how the particular processes work. In the following four chapters each of four broad types of process will be considered in turn. These four are:

1 Cycles – in which outputs become inputs again through a feedback mechanism so producing a cycle
2 Stages – in which outputs provide the inputs for a second set of processes and a progression or series of stages occurs in the evolution of the landscape (Chapter 13)
3 Diffusion – in which the process creates a spreading pattern of outputs through time and outputs are diffused through a landscape (Chapter 14)
4 Events – in which a rapid change in process occurs and extra inputs are suddenly fed into the system with sudden disruption of the current stable process (Chapter 15).

Whilst this fourfold classification does not cover all processes, it serves to show the types of process which operate in the landscape and which can be used to try to explain landscape change. In this chapter the aim is to illustrate some models of cycles. Such cycles can occur within the systems in the natural environment or they can be completely man-made.

The annual energy cycle

The seasonal cycle is essentially an annual energy cycle with inputs of solar energy varying through the year (see also Chapter 9). Fig. 12.1 shows the annual generalised energy cycle of the tropical, mid-latitude and polar regions. Superimposed on the annual cycle is the diurnal cycle. With the cycles in tropical regions the influx of solar energy is high all the year, and also the energy input variation is greater from day to night than it is from winter to summer. The annual and diurnal energy cycles in mid-latitudes show the opposite relationship and energy influx is less. In the polar case the annual cycle dominates, but the intensity of radiation is never very high. The radiation balance for the polar areas is such that more energy is given off by terrestrial radiation than is received from solar radiation. For this state to exist a poleward flow of energy from temperate and tropical regions must occur. Part of this flow has already been discussed in Chapter 9 in the account of ocean currents. The tropical regions receive more radiation than they emit, and in the polar regions the reverse occurs. In order to maintain the earth-space energy balance, transfers occur towards the poles.

Alongside the generalised graphs of Fig. 12.1 are actual graphs of temperature and precipitation for three places. The seasonal cycles of precipitation are shown with peaks at periods of maximum energy input. There are important exceptions to this general rule, with some climates having winter rains. The double peak in Ho Chi Minh City (formerly Saigon) is associated with the twin seasonal peaks of energy input. In the mid-latitude example of Pittsburgh, USA, the precipitation peak is less pronounced. These energy cycles are important underpinnings to a range of cycles in the landscape, not only those related to agricultural production. However, it is in agricultural landscapes that man most clearly adjusts his activities and creates a landscape in response to basic seasonal cycles.

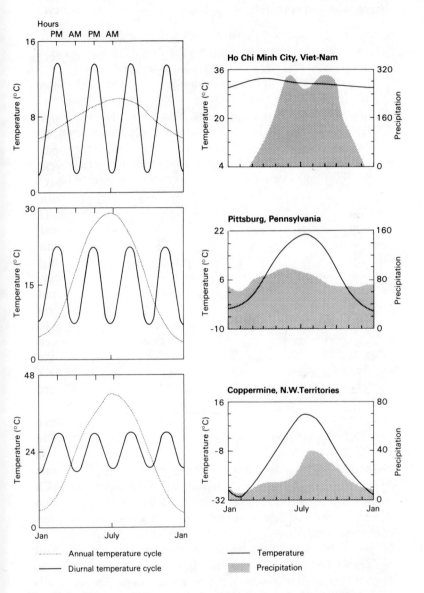

Fig. 12.1 Energy cycles at tropical, mid-latitude and polar locations

Environmental nutrient cycles

Cycles typify many of the processes within stable ecosystems in which various nutrient elements in the biosphere are conserved, used and recycled. Of the elements comprising global living matter, 99.4% is accounted for by hydrogen (49.7%), carbon (24.9%) and oxygen (24.8%). The production of carbohydrates, involving these elements, through photosynthesis is basic to the nutrient cycles in the biosphere. Fig. 12.2 shows the carbon cycle in simplified form. In the carbon cycle, carbon dioxide in the atmosphere is lost to the land through surface transfer and through photosynthesis, but is returned through respiration and the combustion of fossil fuels. In

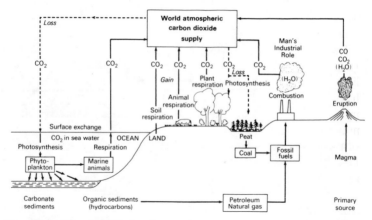

Fig. 12.2   The simplified carbon cycle

the land based part of the cycle the carbon is present as carbohydrates in organic matter, as hydrocarbons in rocks such as coal and oil bearing strata, and as minerals such as calcium carbonate. Man uses the hydrocarbons and injects considerable amounts of carbon dioxide into the atmosphere. Before industrial society developed to any significant extent the amount of carbon dioxide in the atmosphere was constant, but in the last 120 years there has been an increase of about twelve per cent in atmospheric carbon dioxide due to the burning of fossil fuels containing hydrocarbons. The rate of increase of carbon dioxide is itself increasing, giving cause for concern over the environmental effects of these higher concentrations in the atmosphere. An increase in carbon dioxide in the atmosphere causes a rise in the amount of incoming and outgoing radiation absorbed by the atmosphere, so changing the energy balance discussed in Chapter 9. The result would be an increase in earth temperatures. There is no concrete evidence for such a heating phenomenon and some scientists suggest that the effect of increased carbon dioxide in the atmosphere is more than offset by the veiling effect of increased amounts of dust, resulting from industrial activities, collecting in the upper atmosphere. Small particles in the stratosphere absorb and scatter solar radiation and an increase in particle concentration would result in less radiation reaching lower levels of the atmosphere and a consequent cooling of the earth. The effects of increased carbon dioxide concentrations in the lower atmosphere may well be offset by increased particle concentrations in the upper atmosphere.

The carbon dioxide in the atmosphere is used in photosynthesis by land based plants and also, importantly, phytoplankton use dissolved carbon dioxide in the oceans and seas. Phytoplankton are the basic building blocks of food chains in the oceans and carbon in organic and mineral forms is present in marine animals. The various minerals and organic compounds eventually sink to the sea bed, forming carbonate rocks and hydrocarbon sediments. Ultimately the carbon components become released, perhaps after many million years, and the carbon cycle continues.

The oceans and seas provide the largest store of carbon, with an estimated fifty times more carbon dioxide stored in the sea than in the atmosphere. The atmospheric store of carbon, which is available to plants for photosynthesis, is estimated at 700 thousand million tonnes. Live vegetation probably contains around 400 to 500 thousand million tonnes and the growth processes of this vegetation fix, by photosynthesis, around 20 to 30 thousand million tonnes per year. The carbon cycle in ecosystems, particularly those with a large biomass (see Chapter 7) is responsible for moving large amounts of nutrient material.

The carbon cycle is one of several nutrient cycles present in ecosystems. In the oxygen cycle, for example, photosynthesis releases oxygen into the atmosphere. The basic equation of photosynthesis is:

$$6CO_2 + 12H_2O + light \rightarrow C_6H_{12}O_6 + 6H_2O + 6O_2 + heat$$

but this summarises a complex two phase process.

Carbohydrates, oxygen and water are the main products in this process. In the respiration process oxygen is removed from the atmosphere by absorption by animals. The oxidation process is:

$$CH_2O + O_2 \rightarrow CO_2 + H_2O + heat$$

In some ways, therefore, the oxygen cycle complements the carbon cycle in respect of transfers between the atmosphere and biosphere. The oxygen produced by photosynthesis is used in oxidation to reduce carbohydrates back to carbon dioxide and water. It is estimated that the oxygen in the atmosphere is recycled approximately over 2000 years (see Chapter 3).

The nitrogen cycle is based on the vast atmospheric reservoir of nitrogen. Nitrogen moves out of this reservoir via nitrogen fixing bacteria which convert atmospheric nitrogen into more active forms. These can be used directly by plants, which in turn change them into organic forms which are consumed by animals. Nitrogen fixation is an ability possessed by only some micro-organisms. Some soil bacteria are of this type and their presence provides a nitrogen supply to the soil. One group of the decomposer bacteria carry out the reverse process and remove nitrogen from organic compounds and effectively release it into the atmosphere. As changes take place from organic to inorganic and back again so a cycle of nitrogen flows takes place. The nutrient cycles in stable ecosystems are in almost perfect balance. Disruption of the balance creates instability and can cause major changes in the plant and animal components of the ecosystem (see Chapter 16).

Rice growing as a seasonal cycle

The seasonal energy cycle alongside the nutrient cycle together generate a very basic landscape cycle affecting many types of activity. All agricultural landscapes are affected by these two cycles, and their importance may be illustrated by an example. The agricultural system of intensive cultivation of rice in South East Asia has developed over many centuries and is a response to the *physical conditions* of:

Temperature – most rice varieties will not grow if temperatures fall below 20°C–30°C for a minimum of 100 days

Water – precipitation of 1000 mm/year is needed with 125 mm/month in the growing period and a short dry period for ripening and harvesting

Soils – an optimum *pH* range of 5.5 to 6.5 in a soil impermeable enough to prevent excessive moisture loss, such as clays and clay loams on level lands to allow for water management

These near optimal physical conditions for rice may be compared with the physical requirements of other major crops, as shown in Table 12.1.

Rice cultivation in South East Asia is also a response to the *cultural conditions* of:

Population pressure – rice is high yielding, has a high food content and can be stored for relatively long periods

Communal traditions – the presence of highly cohesive social groupings allows strict management of water and concerted group efforts at key times in the production cycle (e.g. harvest)

Religious and political influences – many agricultural practices are associated with particular Hindu, Buddhist and Moslem beliefs whilst governments try to improve agricultural practices and may impose land reforms

Technological levels – manual methods are generally used with a very low level of technological input and high labour input, but exceptions do occur, as for example in Japan.

These physical and cultural factors in combination account for the dominance of wet rice cultivation in monsoon Asia. Whilst the physical conditions allow the growing of rice, the cultural factors are responsible for the farming methods and practices. Although the actual time in the annual cycle when the main farm operations are carried out varies from place to place in response to climate and local tradition, the overall cycle usually involves five successive activities:

1   Preparation of water supply
2   Ploughing
3   Sowing and transplanting
4   After-cultivation
5   Harvesting

Table 12.1 Physical requirements for some major crops

	Growing season (days) and temperature conditions (°C)	*Annual precipitation and distribution (mm)*	*Other environmental conditions*
Wheat	90–100, cool to warm	250–750, abundant during growth, drier when maturing	Rolling to level land, has some drought tolerance
Rice	100+, above 20°	1000 minimum, even on summer maximum	Level land facilitates irrigation, poorly drained soils useful, poor drought tolerance
Millets and sorghums	90, 20°	500+, concentration needed during growth, drier when maturing	Rolling to level land, has some drought tolerance
Corn	140+, 23°, warm days and nights	750–1200, concentration needed during growth, with warm temperatures	Rolling to level land, very adaptable, but not drought tolerant
Soybeans	100+, 23°, warm days and nights	750–1200, concentration during vegetative growth, dry in maturity	Rolling to level land, very adaptable, similar to corn, slightly more drought tolerant
Sugar cane	365, warm to hot	1000+, abundant and even for 11 months	Rolling to level, photoperiod critical (length of day)
Cotton	180–200, 20°+, warm nights	750–1500, summer concentration, dry in maturity	Rolling to level, requires sunshine and regular moisture

Control and management of the water supply is critical to the production of wet, or paddy, rice. Water control may simply involve transferring water from nearby streams, but often there is an intricate system of irrigation canals and ditches. Water may have to be lifted to higher levels and waterwheels and shadufs are used for this purpose. For ploughing, a single furrow plough drawn by a water buffalo is commonly used. This type of wooden plough is cheap and easily produced, but ploughs only to the relatively shallow depth of about 75 mm. The ground is usually ploughed with a water cover and so the timing of ploughing is related to the arrival of the first monsoon rains unless irrigation water is available. Usually the land is left for two or three weeks after ploughing before the seedlings are transplanted. The two stages of land preparation are generally less labour intensive than later parts of the cycle.

The third major activity is sowing and transplanting. Seed may be either sown into the prepared fields or into specially prepared nursery beds and later transplanted into the paddy field. Direct sowing tends to result in lower yields but requires less labour. Transplanting is most widely used in the densely populated areas where there is both plentiful labour and pressure on space in paddy fields because of attempts to produce two rice crops per year. Seedlings are transplanted from heavily fertilised nursery beds after three to four weeks. Replanting is usually done by women and children working communally and the spacing of the rice plants varies between 10 and 30 cm depending on plant variety. The advantage of transplanting lies in the higher yield that is obtained. This is due to being able to time the planting more precisely to suit particular water conditions, so that small seasonal variations can be used to advantage, and to allowing more time for preparation of the paddy field whilst the plants are in the nursery. Planting in rows also makes the after-cultivation of the crop easier.

After-cultivation involves weeding and draining in preparation for the harvest. Weeding is often carried out by hand but some mechanisation has been introduced in recent years and chemical weed killers are also increasingly used. With direct sowing it is difficult to use mechanical cultivation aids and weeding has to be carried out by hand. Harvesting is another manual operation, using sickles or small knives. The rice is then prepared for milling or for storage by threshing and winnowing. This also involves a consider-able amount of communal manual labour.

The five groups of activities typically take around four to five months, with the precise time of starting depending on local climatic conditions. Activity is often concentrated in the May to October wet period, but several methods have been devised for prolonging activity through the seasonal cycle to allow two crops of rice to be produced within a single seasonal cycle. There are three ways of getting two crops from the same land in one seasonal cycle. If the wet growing season lasts for about six months, then it is possible to grow two crops successively as rice requires a minimum of three months growing period. This is possible in the Ganges Delta region where heavy cyclonic rains occur before the main monsoon rains and allow early preparations in March and planting in April. The first crop can then be harvested in September and a second crop can be planted immediately. In Madras in Southern India the monsoon normally breaks early and the same timing of a double crop is possible. Where the wet season is not so long, a double crop can be achieved by sowing seeds of a second crop in nursery beds before the first crop is harvested. Again pre-monsoon cyclonic rains in parts of Central and Southern China allow an early sowing of the first crop, which can then be harvested in August, allowing the transplanting of nursery-grown seedlings in the paddy fields for a second crop harvest in November. Where the growing season is even shorter, two crops are obtained by interplanting, in alternate rows, seedlings of two varieties which take different times to mature. This practice evens out the labour input during the growing season. The early crop may be harvested in July but the second crop, because of its longer growing period and because it has been shaded by the earlier crop, is not harvested until October. This intercropping method is typical of rice growing areas where the growing season is limited by relatively low winter temperatures but water management allows cultivation to begin as soon as tempera-tures permit. These three types of *multiple cropping* allow two rice crops to be grown in the same paddy field within one seasonal cycle.

Rotational cropping occurs where rice is grown during the wet season and a second product is grown during the other part of the cycle. Although the winter season may be too cold or dry to allow a second rice crop, it may be possible to grow sweet potatoes or another vegetable or even a second cereal such as wheat or barley. In much of Japan the summer rice crop is followed by a winter vegetable crop and the pattern is also seen in Malaysia and Taiwan. The non-rice crop may result in increased rice crop yields as soil fertility is not depleted to the same extent as if rice crop followed rice crop. Rotational cropping not only helps conserve soil fertility but also provides a more varied diet for the agricultural population and makes maximum use of the cropland. As with multiple cropping, rotational cropping occurs within a strict seasonal cycle and can only be understood and explained in terms of this cycle.

Market cycles

Ecosystem and seasonal cycles result from processes within the natural environment, but some cycles in the landscape are purely man-made and result from man's organisation of the society in which he lives. One example is the cycle of market activity which takes place in many societies. Markets for food or clothing are often periodic rather than permanent and continuous. A *periodic market* occurs when the market is not open every day but only once every few days on a regular basis. Such markets often occur when the total demand is not sufficient to support extra permanent shops and consumer mobility is relatively low, so traders travel around several markets accumulating enough effective demand to support their businesses. Periodic markets are common in less developed countries where the rural population has a very low spending capacity and personal mobility is very low, but such markets also are a feature of many towns in Europe where on a particular day each week stall traders operate. Periodic markets effectively concentrate demand in specific places on specific days, and from the consumer point of view they reduce the distance consumers need to travel to obtain the goods they require.

Cycles exist in periodic markets as traders move around a sequence of markets in a regular way, returning, after a fixed interval, to the market from which they started. The length of the market cycle may vary from two to ten days, although in most Christian countries a seven day cycle is widespread. In West Africa various tribal differences have led to different lengths of cycle and Fig. 12.3 shows the differences in length of cycle in part of West Africa.

Although the seven day market week is widespread this only seems to have been introduced after the arrival of Islam. The four day cycle is common in Nigeria.

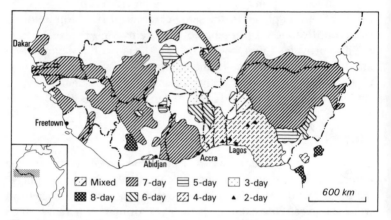

Fig. 12.3   Periodic market cycles in an area of West Africa

Detailed work on market cycles has been carried out in China and a model of the operation of periodic markets has been devised. The basis of the model is shown in Fig. 12.4. There are three levels of market – standard markets, intermediate markets and central markets. Central markets, the largest, are places where imported goods are received and sent out via market traders to the smaller markets. Central markets also serve the reverse function of a collection point for goods to be sent to other central markets or even for export. The market days, within a ten day cycle, are shown on Fig. 12.4. Each market has three market days within a complete cycle. The central market operates on days 1, 4 and 7 and would operate on day 10, but this is a marketless day. Standard markets exist between central and intermediate markets and traders move in sequence from central market to standard market 1, standard market 2 and back to the central market. So intermediate market 1 functions on days 2, 5 and 8 and intermediate market 2 on days 3, 6 and 9. Traders move around a triangular route. The intermediate markets operate in a different sequence of markets but one which overlaps with the circuit based on the central market. The market days in the different circuits are synchronised in such a way that on any given day

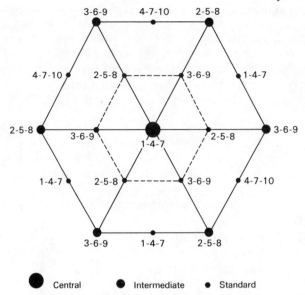

Fig. 12.4 A simple model of the location and operation of a periodic market cycle

adjacent markets are not functioning. This model explains the pattern of markets in China, but has been used in a number of other areas although the actual market day sequence is different with market cycles of different lengths. This integrated pattern of market location and operation is a logical and convenient way of trading in a low income or sparsely populated area.

Such circuits and cycles are not limited to providing goods at market places. Circuit law courts operate on the same principle, with court cases being collected and heard as a judge travels around a precisely defined circuit. Entertainment services may also follow such a cycle, with funfairs, circuses or opera companies moving from place to place, stopping a few days in each place but returning at the end of each cycle to a base. Such a process gives the population of small places access to entertainments otherwise unavailable and allows the entertainment organisations to operate at a profit which would not otherwise be possible. Such cycles, in distinction to those considered earlier in this chapter, are purely man-made.

The cycles of poverty and growth

Processes may themselves operate in such a way as to result in a cycle, with each process generating another but with feedback to an earlier operating process. The cycle of deprivation or circle of poverty which constricts the formation of capital and limits economic growth in less developed countries is of this type. The sequence is:

Low real income in the population
↓
Low buying power and savings capacity
↓
Low rate of capital formation
↓
Lack of capital
↓
Low productivity

For economic development to take place it is necessary to find a way of breaking out of this cycle and moving into an opposite positive cycle which produces growth. The usual way is to try and improve the process of capital formation. Aid from the developed world can be important in breaking out of the cycle of poverty into a cycle of growth.

The positive cycle of development is shown in more detail in Fig. 12.5. The introduction of new industry produces greater employment, which in its turn attracts other industries linked to the new industry. The injection of wages results in generally greater wealth, an increased tax base and the provision of infra-structure to attract yet more new industry. There are several feedbacks and short circuits in this cycle. The effect of the cycle is to notch up the level of economic development each time the cycle operates. Development in this case is seen as a continuous process but other models of development, as we shall see in the next chapter, view it as a process comprising a series of clearly defined stages.

Implicit in this growth cycle is the idea that development hinges on industrial growth, but in some countries this idea has been questioned. Where industrialisation has occurred it has not always been beneficial with, in some cases, new industry creating isolated

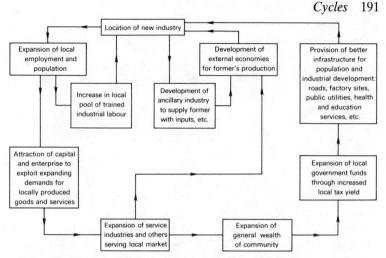

Fig. 12.5 A model of the cycle of economic development

growth centres employing a privileged, industrialised, industrial workforce with a quite different lifestyle from the rest of the population. These inequalities may be perpetuated by government policy, as for example in Brazil where industrial investment by government has been concentrated in the South East of the country at the expense of the densely populated North East. The models of cycles provide a framework and checklist against which to match actual changes occurring in the landscape.

Cycles of the type discussed in this chapter usually represent stable processes which would continue for a long time if left to themselves. Man often intervenes in the process, however, and acts as a disruptive force, sometimes with the best of intentions and for the good of society, as in the case of attempts to break out of the poverty cycle. Sometimes intervention can be harmful and the cycle is disrupted, as for example with various pollutants introduced into stable nutrient cycles in ecosystems. The processes in the cycle may be stable enough to absorb such pollutants but breakdown of the cycle can occur and instability and environmental crises result (see Chapter 14).

13

Stages

Models based on the idea of development stages consider that landscape change progresses through a series of definable stages, each of which clearly differs from what has gone before. Within each stage the processes of landscape evolution are broadly similar, but the processes differ substantially from one stage to the next. Stages may be considered as staged over time, or alternatively staged over space, with a sequence of landscapes existing at any one time.

Stage models of economic development

Several of the models of economic development follow the idea of a progression through stages. Marx suggested four different historical stages of economic and social development and two future stages. At a critical point in each stage internal conflict led to the breakdown of the processes of the older stage and the emergence of new processes and a new stage of society. In each stage he argued that the agents of change were the social classes created by a particular mode of production. The earliest stage of *primitive communism* broke down with the domestication of animals and plants. At the same time the development of primitive technology for metal smelting led to the production of surpluses in some places and the specialisation and organisation of labour. One social group organising the labour resulted in the second stage – *slavery*. Slavery was replaced by *feudalism* as labour become increasingly specialised, technology improved, a merchant group emerged and there was need of controls over the increasingly complex society. The fourth historical stage is *capitalism* in which the worker does not own the methods of production, but sells his labour to capitalists who own the means and methods of production. In the capitalist stage, decisions about production become concentrated in the hands of a very small number of people and Marx argued that conflict between decision makers and workers would result in a fifth stage of *socialism*. This in turn would collapse in favour of a final stage of *communism*. In this model the basis of landscape change and economic growth is technological change creating conflicts among groups in society. In resolving these conflicts society itself changes and economic development occurs.

An alternative model of five stages of economic growth has been suggested by an American economist, W. Rostow. He argues that the critical factor which moves countries from one stage to the next is not social conflict but capital availability and use. The five stages are:

1 The *traditional* society in which most workers are in agriculture, levels of technology are low and age-old production methods are used. There is acute capital shortage and society is locked into the poverty cycle discussed in the last chapter.

2 A *transitional* society in which the population becomes aware of the benefits of economic growth, contacts with other countries increase and government policies are introduced to encourage economic growth and development. Changes in attitude occur and some small attempts are made to introduce industry.

Societies in these first two stages are typically dominated by agricultural activities and by small scale cottage industries, such as weaving or making small scale agricultural implements. The rural villages of the less developed countries of Africa, South East Asia and India are examples of societies at these first two stages.

3 A period of *take-off* when the rate of economic growth quickens and industrialisation takes place. Personal incomes begin to rise as capital investment takes place and the transition from the poverty cycle to the growth cycle occurs. The capital is used to improve technology and so to raise productivity. Investment tends to be urban based and the rural–urban differential in opportunity leads to the initiation of rural–urban migration of the population. This in turn results in a reduction in the political

and social power of rural landowners in favour of the new employers in urban based factories. The take-off period can be quite short and effectively last for only twenty or thirty years, but in other cases it can be slower and subject to fits and starts as fluctuations occur in the amount of capital available for investment. Table 13.1 shows some dates of the take-off stage in different countries. The precise time that take-off occurs, according to Rostow, depends on the critical rate of capital investment.

Table 13.1   Approximate dates of economic 'take off' of selected countries

Great Britain	1783–1802
France	1830–1860
Belgium	1833–1860
USA	1843–1860
Germany	1850–1873
Sweden	1868–1890
Japan	1878–1900
Russia	1890–1914
Canada	1896–1914
Argentina	1935–
India	1952–

4   A stage of *maturity* in which widespread rises occur in living standards, urbanisation and industrialisation. Innovation and new technologies are no longer limited to industry but are applied to agriculture, which operates more productively but with fewer workers than in previous stages. This stage is a period of intensification of the processes which characterised take-off. Factories expand and economies of scale are exploited: production costs for an individual article can be reduced by producing the article in very large quantities. Most industry is based on local agricultural production or local physical resources, and landscape changes with the emergence of small factory towns and improved transport networks linking these developed nodes.

5   A society of *high mass consumption* is the final stage in the Rostow model. Personal incomes are high and there is an abundance of goods and services with portions of the population

no longer concerned with securing the basic necessities of life. Industry takes increasing account of economies of agglomeration whereby economic efficiency is improved by a range of industries locating close together and linking themselves together in a complex pattern of production. These economies of agglomeration lead to the formation of large urban–industrial regions such as exist in Europe and North America. In these societies there is a steady increase in expenditure on services and leisure activities.

The five stage model of economic development envisaged by Rostow ends with a society and economy of high mass consumption. Other people have suggested a further stage which has been called *post-industrial* society by some and *post-materialistic* society by others. In this stage the high levels of capital investment already achieved cease to increase and sections of the population react against ever-increasing production and the high levels of consumption. As might be expected this reaction comes from wealthy societies in which consumption is high, such as in parts of suburban West Germany and in California. It is envisaged that the landscape of this sixth stage will be characterised by small groups engaged in communal activities of self-sufficiency in rural areas and a totally decentralised city pattern as micro-electronic technology removes the need for people to travel to cities to work and allows them to work in their own home.

There are other models of economic development which have been suggested by geographers and economists but most argue for the presence of specific stages of development. The models differ most in what they consider to be the main reason for transition from one stage to the next.

The Von Thünen model of agricultural land use

The large scale international differences in the economic landscape have been modelled by ideas of development stages, but of a temporal sequence of stages. On a smaller scale the various agricultural land use patterns around a city may be seen as representing different stages, with each stage characterised by specific land use and sharp transitions from one stage to the next. These are a spatial sequence of stages, however, not a temporal one. A model of agricultural land use zones around the city was developed by a

German landowner, Von Thünen, early last century, but the model has been extended considerably by geographers, particularly since the early 1950s. The basic model suggests that agriculture around a city will be geared to the market for agricultural goods which exists in the city. Distance from this city market will be critical in influencing what products are grown, as each product grown at a particular place will have associated with it a transport cost in order to get it to the city. Products from further away will have higher transport costs. The price at the city will be a combination of production cost and transport cost. Initially, for simplicity, it is assumed in the model that the production cost for a particular product is the same wherever it is produced. In the following example imaginary values are taken for the various distance and profit measures to illustrate how the model works. For a particular product, for example milk, we can show the transport cost and price at the market on a graph, as in Fig. 13.1, which shows production costs fixed at £30 per hectare. Here the market is at A. In order to meet the demand for milk at A all the farmers within a distance of 7 km from A must produce milk. The people in the city at A have to pay a price for milk which allows the farmer 7 km away to break even. The sloping line then becomes the transport cost line for milk because we have assumed production costs to be the same everywhere. The farmer at 7 km is the *marginal producer* and the sloping line is called the *transport gradient* for milk. It costs the marginal farmer £70 per hectare to transport milk to the market. If we consider a second farmer at 4 km, his transport gradient is the same but it only costs him £x to transport milk to the market. So the second farmer gets a profit of $70 - x$ per hectare, shown as Y on the graph. This amount Y (£30), is the profit or extra money the second farmer gets from his saving on transport costs by being at a place 4 km from town rather than a place 7 km away. This profit Y is called the *economic rent* at a place and is clearly related to the distance of the place from the market. A farmer more than 7 km away from the town would make no profit if he produced milk, and so he would grow a crop with a different, but shallower, transport gradient to produce a profit for him. The transport gradient line is also the line of economic rent or *net profit*.

With only one crop the graph is relatively simple, but if three crops are introduced it becomes more complicated, as shown in Fig. 13.2. The transport gradients for dairy products, grain and meat are

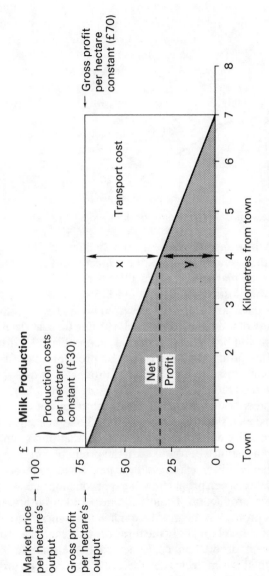

Fig. 13.1 Imaginary transport costs and market price for milk in the Von Thünen model

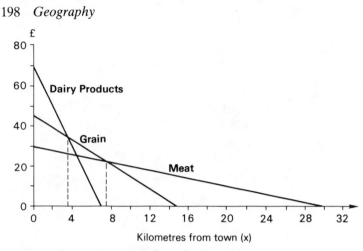

Fig. 13.2   Economic rent gradients for three crops

shown, with the transport gradient being shallowest for meat. Table
13.2 shows the assumed values of the critical measures of the three
operations. The market prices are £100, £65 and £45 per hectare,
and production costs are £30, £20 and £15 respectively. A farmer
operating only 2 km from the town could make a profit from
producing any of the three products and selling them at the market,
but the profit will be highest, as can be seen in Fig. 13.2, for dairy
products. This product will therefore be produced closest to the
city. Transport costs are high for dairy products, so the transport
gradient is steep (£10/kilometre/hectare – see Table 13.2), and net
profit falls steeply with increasing distance from the city. In Fig.
13.2, at a distance slightly less than 4 km (3.6 km exactly) the net
profit from a dairy enterprise is the same as that of a grain enterprise and for a distance beyond 3.6km the profitable product would
be grain. This condition exists until the next stage is reached at 7.5
km from the city, when profits from meat production become higher
than for grain production. From 7.5 km to 15 km both grain and
meat yield a profit, but the profit from meat is higher. After 15 km
only meat is profitable at all. In the three product examples shown in
Fig. 13.2 meat production continues to the economic margin.
According to this simple model there are three zones around a
town, each with a characteristic land use and with profits from each
land use decreasing with distance from the market in the town.

Table 13.2 Imaginary returns and costs per hectare for farmland in three activities

(1) Crop or land use	(2) Market price per hectare of output per year	(3) Production costs per hectare of output per year	(4) Gross profit per hectare of output per year (Col 2 – Col 3)	(5) Transport rate per kilometre per hectare of output	(6) Margin of profitable production: kilometres from town (Col 4/Col 5)
Dairy	£100	£30	£70	£10	7 km
Grain	£65	£20	£45	£3	15 km
Meat	£45	£15	£30	£1	30 km
Any crop	x	y	$x - y$	z	$\dfrac{x - y}{z}$

Changes in land use occur with distance from the town with first a stage of dairy production, followed by a stage of grain production, followed by a stage of meat production.

The critical points are the distances from the town when one stage or type of agricultural production changes to another. These distances can be calculated as the points when two land uses give the same economic rent or net profit. The economic rent is calculated as:

$$R = (X - Y) - ZK$$

assuming that the yield per hectare is constant:
where X is market price per unit of commodity

Y is production cost per unit of commodity

Z is transport rate per unit per km

K is distance in km of the point of production from the market

In the example in Fig. 13.2 and Table 13.2 the economic rent or net profit for a dairy farm at 3 km from the town is:

$$R = (100-30) - (10 \times 3)$$
$$= 70 - 30$$
$$= £40 \text{ per hectare}$$

For dairying $R = £70 - 10K$
For grain $R = £45 - 3K$

When the economic rent for the two crops is the same, then land use changes. So when

$$70 - 10K = 45 - 3K$$
$$\text{i.e. } 25 = 7K$$
$$\text{i.e. } 3.6 = K$$

then land use changes. The outer edge of the dairy zone is therefore 3.6 km from the city.

The model as it has been worked through with the three crop example is over-simplified. It could now be extended to take into account changes through time in the market price, production costs and transport costs, so developing a model of land use change. The essential pattern remains of land use changing with distance from a town. It is unusual for very obvious zones of land use to exist around a city as there are several more factors determining actual land uses than the few considered in the simple model. The model serves as a first step in attempts to explain agricultural land use variation by showing the importance of transport costs changing through the landscape. Whilst the models of economic development suggest stages of landscape evolution through time, Von Thünen type analysis shows how stages of agricultural land use occur through space.

Land forms of mountain glaciations

Models which seek to explain how a landscape passes through various evolutionary stages are not limited to economic geography. In explaining the physical features of the landscape such models are also in widespread use. This is particularly so in the study of land forms. Sequences of stages of land form development are identified in coastal landscapes or in arid landscapes. The following example shows the stages seen in glaciated landscapes. The stages are related first to temporal considerations of the sequence of successive ice ages and then to the spatial sequence of glacial land forms within a glaciated landscape. An important division in glacial land forms is between features resulting from erosion and those from deposition but the majority of features we see in present day landscapes are *relict* features formed during earlier glaciations. The last major ice sheet over Europe disappeared around 9000 BC and over continental Canada around 4000 BC. These last glaciations were the most recent of a series of probably at least ten major periods of northern hemisphere glaciation in the last million years. The cold periods probably lasted for around 50000 years, but within the cold and relatively warmer periods there have been minor periods of cold and warm which only lasted a few hundred years but caused mountain glaciers, rather than ice sheets, to expand and retreat. Table 13.3 shows the probable pattern of temperature and glacial trends in the northern hemisphere in the last 700000 years. During the middle and late Pleistocene there were sequence glacial and interglacial periods. In different parts of the northern hemisphere these glacial stages have different names and Table 13.4 shows these. Before about 700000 BP, in the early Pleistocene, there is really only conjecture about the number of glaciations in the

Table 13.3 Glacier and temperature trends in last 700 000 years

	Dates	World Trend
	Since AD 1895	Minor warming; mountain glaciers retreating
	AD 1500–1895	'Little Ice Age'; mountain glaciers readvance
	AD 1000–1300	Warm spell; mountain glaciers retreat
Holocene	1000 BC	Major cooling trend begins; mountain glaciers readvance
	3000 BC	Warming trend begins; mountain glaciers retreat
	4000 BC	Major cooling trend begins; mountain glaciers readvance
	5000 BC	Maximum recent warmth; final disappearance of continental glacier in Canada before 4000 BC
	8300 BC	Major warming trend begins, with disappearance of continental glacier in Scandinavia about 7000 BC. Temporary readvances of mountain glaciers about 8000 and again 6000 BC. Sea level approximately 40 m below current levels.
Late Pleistocene	75 000 BP	Last glacial begins. Formation of new continental glaciers in Canada and Scandinavia. Expansion into the United States and across the Baltic Sea, by 60 000 BP. Limited de-glaciation after 40 000 BP was followed by maximum glaciation (Newer Drift) about 25 000 BP. The North American glacier reached south to latitude 39°N, the European to 52°N. Rapid de-glaciation 15 000–11 000 BP.
	125 000 BP	Last interglacial begins. Climate and vegetation similar to that characteristic during Holocene, interrupted by colder intervals with expanded glaciers about 115 000 and again 95 000 BP.
Middle Pleistocene	700 000–125 000 BP	Alternative glacials and interglacials. Probably at least four intervals of major, northern hemisphere glaciation (Older Drift) that brought severe cold to mid-latitudes.

Table 13.4 Glacial and inter-glacial stages in the northern hemisphere since the early Pleistocene

Provisional numerical order	Great Britain	Alps	Northern Europe	European Russia	North America
Last glaciation	Devensian	Wurm	Weichselian	Valdai	Wisconsinan
Last interglacial	Ipswichian		Eemian		Sangamonian
Fourth glaciation	Wolstonian	Riss	Saale	Dnepr	Illionian
Third interglacial	Hoxnian		Holstein		Yarmouthian
Third glacial	Anglian	Mindel	Elster	Likhvin	Kansan
Second interglacial	Cromerian				Aftonian
Second glacial		Gunz	Pre-Elster	Pre-Likhvin	Nebraskan
First glaciation		Donau			Pre-Nebraskan

Source: Table 8.23 in M. J. Bradshaw, A. J. Abbott and A. P. Gelsthorpe, *The Earth's Changing Surface*, Hodder and Stoughton.

preceding two million years. Most of the present day glacial features date from the last two or three glaciations, and so in geological time are relatively recent features.

Mountain glaciers result from the gradual accumulation of snow in perennial snow patches in *nivatian* hollows. Snow collects in these slight depressions and the air spaces are eliminated through compaction by the snow's own weight and through a melting and refreezing process. As the air pockets disappear the new material is called *firn* and the third stage, as recrystallisation occurs, is ice.

Ice is quite different from snow because ice, under pressure, can be deformed and be subject to *plastic flow*. When ice thickness reaches around 30 to 60 m and flow occurs, then it becomes a *glacier*. The rate of glacier flow depends on ice thickness, ice temperature and on the slope and bedrock character. The greater the ice mass the faster the flow; ice close to freezing point flows faster than very cold ice. Small cold glaciers on shallow slopes may only move a few metres a year, but larger glaciers may move several hundred metres per year and there is evidence of some glaciers in Greenland flowing several kilometres per year. There are several stages of development, therefore, between a snow patch and a glacier. In particular highland areas there is normally a balance between snow accumulation and snow melting. Above a certain critical limit, the snow-line, more snow falls than melts during the summer and so perennial snowfields and glaciers develop. The combined losses from melting and evaporation are termed *ablation*. Below this limit, snow that falls during winter melts during summer, but glaciers from above the limit may extend below it as they are supplied with ice from the high snowfields. The end of the glacier, its snout, usually extends during winter and contracts during summer, but this seasonal cycle may be superimposed on a long period sequence of change involving long term glacier advance with a lowering of the snowline, or glacier retreat. As suggested in Table 13.3, glacier retreat is currently taking place in the northern hemisphere.

When glacial retreat is completed then a new landscape is exposed. Features resulting from glacial activity become relict features and begin to be modified by other agents of landscape change. A number of landscape features characterise highland areas where valley glaciers have been present.

Cirques are amphitheatre shaped depressions, cut in bedrock, which occur at the head of the valley and were the source of the valley glacier. They comprise 1) a basin with smoothly sloping floor, commonly with a swampy area or small lake (tarn) in the centre; 2) a steep, near vertical backwall which may be as high as 1000 m and often has a notable lack of debris at its foot, suggesting that during its formation glacial processes removed this rock waste; 3) a low rock threshold at the exit of the basin and separating the cirque from the valley below. Conditions which seem to favour maximum cirque development are the rather wide spacing of valleys, valley glaciers with large snowfields, and a homogeneous rock susceptible to the plucking action of ice. Headward erosion of cirques often means that they join, together undercutting mountain peaks from several sides. When this happens sharp-edged ridges (*arêtes*) separate the cirques and the original mountain peak is reduced to a steep sided *horn*.

Glacial troughs are the pre-glacial stream-cut valleys which have been widened and straightened by the erosive power of the glacier. The original spurs of land in the former river valley have been truncated as the trough was formed. The steep *trough head wall* separates the trough from its cirque. The trough is U-shaped in cross section, with steep valley sides and a comparatively flat valley floor. The long section of the trough shows irregularities as glacial erosion has eroded softer rocks to a deeper level than the harder ones. Sometimes the depressions become lake filled, and the floor of the valley contains a series of small lakes (paternoster lakes). In some cases a single larger lake (finger lake) occurs if the mouth of the valley becomes blocked. The glacial troughs therefore descend in a series of steps with the rock bars called *riegels*. Tributary small troughs joining the main glacial trough have been filled with smaller glaciers with less erosive power than the main valley glacier and so the side valleys have been cut less deeply than the main valley. As the ice retreats these tributary valleys are left high above the main valley and are termed *hanging valleys*.

Roches moutonnée are a third characteristic erosional feature of highland glaciated areas. They are rocky outcrops which are rounded on one side and irregularly shaped on the other and they result from the glacier moving over the rock outcrop, smoothing the upvalley (stoss) side and plucking at the downvalley side.

Within glaciated mountain areas there are limited features created by depositional, rather than erosional, processes and the major ones are those created by *moraine*. Any material eroded and transported by a glacier is called *drift* and this is usually divided into *till*, deposited by the ice itself, and *outwash*, resulting from deposition by meltwater beyond the limit of the ice. Moraines are composed of till. Till may accumulate along the edges of a valley glacier, creating *lateral moraines*, at the glacier snout, creating *terminal moraines*, in the centre of a glacier, produced by converging lateral moraines, creating *medial* or *interlobate moraines* or beneath a glacier, creating *ground moraines*. On glacial retreat moraines are left as till deposits in the glacial trough. The thickness of the moraines usually increases downvalley and end moraines which might mark the stable position of a valley glacier over several decades can be 50 m or more high. The other types of morainic features are usually less conspicuous, such as a series of small hills along the edge of the glacial trough or down its length.

Features of lowland glaciation

The more extensive features of glacial deposition occur in lowland areas and comprise till and outwash features. Glaciated lowland areas have often been covered with a sheet of ice rather than by distinct valley glaciers. The edge of the ice sheet, if stationary for several centuries, is marked by the accumulation of a large end moraine which may be well over 100 m high. Features at the edge of an ice sheet are shown schematically on Fig. 13.3. *Till plains* occur in the peripheral zone of a former glaciation and comprise a covering of ground moraine on which other features have been deposited. *Outwash plains* lie beyond the maximum extent of the ice sheet marked by the outermost terminal moraine and are composed of clays, sands and gravels deposited by meltwater from the ice sheet. As the ice sheet retreats and melts, so the till plains are covered with outwash material. Although the features are comprised of outwash they remain classified as till plains.

In till plains the pre-glacial landscape is smoothed and partially buried with drift as valleys are filled and hills masked by deposits. Till plains cover large areas of Southern Canada and the North Central USA, North Central Europe and Northern Russia. Gener-

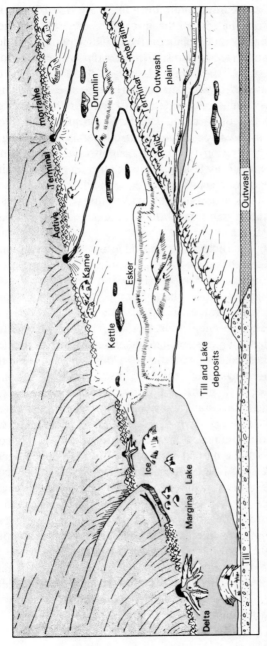

Fig. 13.3 The range of features associated with the stages of de-glaciation of a lowland area

ally in these areas ground moraine is between 25 and 75 m and relief is seldom more than 50 m. Swamplands and poorly drained areas are commonplace on till plains as the drainage pattern has had only a relatively short time to adjust to the new physical landscape. Often ponds and small lakes are widespread: these are places where isolated masses of ice were left behind by the retreating glaciers and melted slowly but at a distance from the edge of the ice sheet. Such small ponds are often called *kettles*. Unsystematic drainage and kettles are particularly common in the till plains produced from the most recent glaciation in what is termed Newer Drift. The most recent glaciation was less extensive than older ones and the till plains of the older glaciations are old enough for systematic drainage patterns to have developed. On till plains a number of particular till and outwash features develop in sequence as the ice sheet retreats in stages.

In Fig 13.3 the relict terminal moraine marks the maximum extent of the ice sheet, the active terminal moraine is being constructed at the edge of the ice sheet, and the recessional moraine marks a place where the glacier retreat halted for a time, perhaps for some centuries. These moraines are till features similar to the moraines in valley glaciers. Terminal moraines may be only a few hundred metres in length or may stretch for many kilometres, and their size is related to the length of time the ice front stayed in one position. *Drumlins* are streamlined, elongated low hills that occur in groups on the surface of till plains. The end of the drumlin pointing to the former ice margin is blunt and steep, whilst the downstream end is shallower and more tail-like. They are orientated to the direction of ice movement and may be as high as 40 m and in length they range from barely 100 m to 2 km or so. Sometimes drumlins have a rock core, but often they consist entirely of till and are caused by the reshaping of ground moraine as ice flows over it. Also shown on Fig. 13.3 are *eskers*, which are steep-sided curving ridges of sand and gravel which can run for 150 km or more. These features mark the former beds of streams which moved through or beneath the ice. Eskers may run uphill as well as downhill, as the streams in the glacier would be flowing under pressure, so allowing them to flow uphill. Subglacial streams may also create small, local deltas, called *delta kames*, where they have emerged from the ice margin straight into a lake ponded up against the ice. These deltaic features are made of sands and gravels which in recent history have proved a valuable source of building material. During each stage in the retreat of an ice sheet these depositional features are revealed as the ice margin retreats. The various stages involved in glacial processes or erosion and deposition produce landscapes with a series of characteristic features.

Stage models of contrasting length

There are many other examples of stage models of use to geographers. Some involve very long periods of time, such as the four stages involving mountain building activity put forward by geologists. They last at least 100 million years, and frequently much longer. The four stages are:

1 A period of deposition of sediment in a geosyncline or large oceanic basin
2 A time of intense earth movements involving folding, faulting and igneous activity associated with the movement of tectonic plates (see Chapter 3)
3 A long period of stability and erosion
4 A final stage of vertical uplift, again associated with plate movements.

Other staged models are of shorter time duration. That of demographic transition, for example, although again based on four stages, takes only several centuries to complete.

Stage 1 is associated with an agriculturally based society where birth and death rates are high and there is a stable or slowly growing population. Large families are common because life expectancy is low and even young children can become members of the workforce. Employment opportunities are few, other than those in agriculture. Stage 2 begins as death rates fall due to better sanitation and improved medical services. Often this is coupled with industrialisation and urbanisation, but the result is a rapid increase in population (see Chapter 8). In stage 3 an urban industrial society is the norm, and there is less need of large families. Children in an urban environment may be an economic liability rather than the asset they were in stage 1 agricultural societies. Birth rates therefore begin to fall and death rates continue to decline but much less rapidly than in stage 2. Population continues to increase but less quickly. Finally stage 4 is a period of low birth and death rates with a generally stable population total, although short periods of increase and decrease will occur in response to short period economic fluctuations.

There are obvious parallels between this model of population change and the models of economic development presented at the start of this chapter. In all such stage based models there are critical points in the overall process of change at which individual processes themselves change and a new mix of inter-related processes characterise the next stage of development. For the sake of simplicity, a distinction between cycle and stage models has been made in this and the previous chapter, but the distinction is not always very clear. Sometimes clear stages may be seen, yet stages themselves, in sequence, may comprise a cycle which can be repeated. Many of the real processes of landscape change are very complex and are certainly far from being fully understood. Nevertheless, when actual changes in a real landscape are measured and mapped using geographical techniques, their comparison with assumed changes expected from a model can be a considerable help towards explaining how and why landscapes change.

14

Diffusion

Change does not occur over all the landscape at the same time. Changes frequently spread through the landscape, changing a specific feature gradually and perhaps leaving others untouched. One process by which a new idea, a particular feature, or even a disease spreads through the landscape over time is called *spatial diffusion*. Several models exist of this process. Diffusion models are particularly applicable to man-made changes in the landscape in which man makes a decision to act, for example to buy a particular piece of new agricultural machinery, and this action results in some sort of landscape change. The individual in this example first gets to know about the existence of the new machinery (exposure to it), secondly he decided to buy it (adoption of it) and thirdly its application results in some form of landscape change. Diffusion models generally aim to understand the first and second of the three states; how information, technology or disease spreads through space so that individuals are exposed to it, and how the individual reacts when so exposed.

Types of diffusion

The spread of an idea, information, or disease is made by *carriers* and limited by *barriers*. Diffusion processes may be divided into two groups depending on whether the number of carriers increases or not. With *expansion diffusion* the number of carriers increases, for example, in the spread of a rumour, one person tells another who in turn tells another and so on. The number of tellers increases and the rumour spreads. Alternatively, with *relocation diffusion*, the movement of people is responsible for the movement of ideas or information. The diffusion of a language may be the result of the relocation of speakers of the language as much as of the teaching of the language to new speakers. In the settlement of a newly discovered territory, settlement diffusion occurs and homes are relocated to new places in the land to be settled. The movement of population in such a case is relocation diffusion, as their number does not increase but their location changes. Expansion diffusion may be further subdivided into *contagious diffusion* and *hierarchical diffusion*. In contagious diffusion, as the name suggests, direct contact must be made between people for the idea or disease to spread. Teller and receiver must be in contact, and often the likelihood of contact depends on how close together they live or work or generally carry on their life. The closer together their lifespaces then the more likely they will be to make contact and be subjected to contagious diffusion. In the spread of the idea to buy a colour television set, for example, a teller (someone with a set) is more likely to show their TV to a receiver (someone without colour TV) if they live in adjacent houses than if they live in different parts of the town. Distance apart of teller and receiver, ie space, is an important factor in contagious diffusion models but it need not necessarily be the only factor – teller and receiver, for example, may be relatives or may work together. Generally, however, in expansion diffusion the strength of the link between teller and receiver is inversely related to the length of the link. The chance of a teller meeting a receiver who lives 50 km away is much less than meeting one who lives 5 km away. Space or distance effectively acts as a barrier to diffusion taking place.

The second type of expansion diffusion is hierarchical diffusion. In this type large places or important people tend to get the information first, transmitting it to others lower down some ordered hierarchy. Information or ideas thus trickle down a hierarchy. Western clothing fashions provide an example of this type of diffusion. Frequently originating in Paris, London, New York or Rome, new fashions are taken up first in the large regional centres of Europe and North America. Subsequently they arrive in the shops of less important cities and finally in small town shops. Even in this strong hierarchical diffusion there is still a strong distance constraint with, for example, small towns in South-East England, relatively close to the major information source of London, being likely to have the fashions sooner than small towns in Northern England. In hierarchical diffusion of this type direct contact may not be necessary between teller and receiver, but the mass media through newspapers and television may be important carriers of information. The purpose of the mass media is effectively to reduce the constraints and barriers imposed by the distance between teller and receiver.

Barriers to diffusion

Distance is not the only type of barrier to diffusion. Other barriers are termed *absorbing*, *reflecting* and *permeable*.

1 *Absorbing* barriers effectively halt any diffusion which comes into contact with them. Impenetrable swamps or uncrossable mountain ranges are physical barriers of this type. The boundary between two aggressive societies who are unable to communicate with each other constitutes an absorbing barrier for the movement of ideas. The boundary of a region in which all the population has been protected against a specific disease is an absorbing barrier for the diffusion of that particular disease. In general such barriers occur only rarely, but when they do, they become important dividers of landscape types.

2 A more common barrier form is termed *reflecting*. In this case the direction of the diffusion process is changed on impact with the barrier. A lake shore, for example, often reflects an approaching diffusion causing it to spread along the lake shore and effectively intensifying its operation along the shore.

3 The third type of barrier is by far the most common and is the *permeable* barrier. On encountering the barrier the diffusion is allowed to continue but at a reduced rate of intensity. An international boundary or a river can often act in this way in the diffusion of an economic or social phenomenon.

Barriers themselves may change their form with the intensity of the diffusion process. A large water body, for example, can act first as a reflecting barrier and then as the intensity of the diffusion process builds up on its shore the barrier may become permeable and diffusion continue on the opposite lake shore. Alternatively diferent permeable barriers may have different degrees of permeability. In the spread of information across international boundaries information diffusion from a place is often measured by the quantity of telephone calls leaving the place. The amount of telephone calls between places divided by a national boundary is less than between comparably spaced places both in the same country. The stronger the political differences on either side of the national boundary the less the diffusion across the boundary. Thus within Western Europe the barrier effect of international boundaries is relatively small, but between a country in Western Europe and one in Eastern Europe the effect will be considerable. Telephone flows between places in Western Germany and France are many times greater than between comparably spaced places of similar size in West and East Germany. The permeability of these various national boundaries is therefore quite different.

A measure of permeability of a barrier is the size of 'normal space' which has a comparable barrier effect, for it will be remembered that diffusion intensity declines with increasing distance apart of teller and receiver. In a study of the Canada-USA national boundary and the Ontario-Quebec provincial boundary within Canada, it was found that the provincial boundary had a barrier effect five times stronger than simple distance, whilst for the international boundary it was fifty times stronger. With other boundaries between more contrasting countries the barrier effect becomes very considerable. Political boundaries are only one type of permeable barrier. Other examples are 1) *physical* barriers of deserts, swamps and oceans; 2) *cultural* barriers, such as strongly defined social classes or religious groupings which inhibit communication amongst people; 3) *linguistic* barriers which may be barriers to communication and hence diffusion. Such barriers are all permeable to different degrees, with some approaching the absorption type barrier whilst others are only minor hurdles in the 'normal' decline of diffusion intensity through space.

Simple space itself may be considered a permeable barrier, with long distance teller and receiver interaction relatively unlikely. As the distance of separation increases, the *probability* of interaction decreases. Each person is at the centre of a *communication field* or *information field* which is strong close to him, but weakens with distance. The decline with distance of the information field will vary from society to society. In very simple societies the decline will be rapid, indicating the lack of long distance communication and the overriding importance of short distance links. In mobile and technologically advanced societies the decline with distance, sometimes called the decay of the field, will be much less. For a particular society it is possible to work out the probabilities of distance decline in teller–receiver interaction. Table 14.1 shows a typical grid of such probabilities, used in North America, in which a teller is imagined to be located in the centre of the table and the probabilities decline with distance from the centre. Thus there is a 4431/10000 chance of a teller–receiver link occurring within the centre square, but only a 96/10000 chance of such a link between the central teller and a receiver from a corner square. In this sample case the probability pattern is regular and the information field is regular around the teller. In reality, of course, the pattern is seldom so simple and a more complex set of grid values has to be used. The presence of differences in population density in the information field and the existence of a variety of barriers to diffusion will result in more complex information fields. These vagaries can be incorporated into the grid of probability scores and a realistic model can be developed to describe and forecast the diffusion process.

Table 14.1 A simplified probability grid table for calculating the probability of contact in a diffusion model

.0096	.0140	.0168	.0140	.0096
.0140	.0301	.0547	.0301	.0140
.0168	.0547	.4431	.0547	.0168
.0140	.0301	.0547	.0301	.0140
.0096	.0140	.0168	.0140	.0096

Such an approach has been used to study the diffusion of irrigation methods to farms in the Northern High Plains of Colorado in the USA. The area is one of dryland farming and cattle grazing. During the late 1940s attempts were made to expand cattle herds, but a series of droughts provided severe problems. A few farms introduced irrigation schemes, so that by 1948 forty-one wells had been sunk and pumps installed to tap the ground water. Whilst the possibilities of pump irrigation were widely known in the area the actual decision to install a pump appears to have been taken only as a result of direct contact with a farm already using the method. The information field of farmers using irrigation, i.e. tellers, was instrumental in determining in which order other farmers, i.e. receivers, adopted irrigation methods. The information field was calculated by a study of barbecue attendance patterns and an analysis of local telephone calls. A probability grid was calculated and so it was possible initially to model what the expected 1962 pattern of irrigation would be, having started from the forty-one wells of 1948. This calculated 1962 pattern was then compared with the actual pattern for that year and the two patterns were seen to be very similar. The study then went on to forecast the post-1962 pattern using the same method of information fields and diffusion to explain this particular aspect of landscape change.

The diffusion curve

The diffusion of ideas and information characteristically begins slowly, the acceptance of the idea becomes more rapid and then slows again as total acceptance approaches. If the percentage of the population accepting an idea is graphed against time then an *f* shaped curve is plotted. This curve is sometimes called an S curve. The reason for the graph taking this form is the way people react to new ideas: some people are quick to adopt new ideas, others are more conservative and some refuse to accept new things until the last possible moment. A group of receivers of a new idea can be divided into four subgroups called:

Innovators who adopt first, are receptive to new ideas and are willing to try out new things. This group is usually about one sixth of the population and several studies in Europe and the USA have suggested that they tend to be younger and have a relatively high income for their age.

Early majority who constitute about one third of the population and who, although not responding immediately to new information, come to a decision to accept the idea more quickly than the third group.

Late majority who again constitute about one third of the population but who are relatively slow to accept change.

Laggards who do not accept the change until almost everyone else has and the change has been proven to be a benefit. Studies suggest that this group tends to be older or poorer than the majority of the population.

Although there is some connection with age and income in this fourfold division, the psychological make-up of the individual is much more important in determining whether he or she is an innovator or laggard. The slow growth of most innovation accept-ance is due to the relatively small number of innovators, and the rapid middle phase of acceptance is due to the acceptance by the early and late majority groups. The psychological make-up of potential receivers is an important factor in determining whether a new idea will be adopted or not.

This psychological factor is one of several which are used to explain the process of decision-making of the receiver once the teller–receiver link has been established. Psychology is particularly important in the diffusion of ideas or of technology use, but is less important in the diffusion of disease, where the receiver may have limited control over whether he or she 'adopts' the disease. The reaction of an individual to being exposed to an idea depends upon the assessment by the individual of the usefulness of the idea. The individual may consider the financial implications of adoption and assess these alongside the potential benefits. How much is it going to cost to sink a well, using the irrigation example discussed earlier, and how soon will these costs be recouped through larger herds of cattle? Having decided that this might be x years, then the indi-vidual will decide whether that investment is worthwhile. Some farmers, innovators and early majority, will decide that it is worth-while fairly quickly whilst others will hold out longer against making a decision in favour of investment. With disease the reaction of the individual is obviously very different and may depend upon factors outside his or her control. The spread of disease such as measles through a school is a process of teller and receiver making contact. The reactions of the receivers depend on whether they have had the disease before or whether they are immune to it in some other way and also on several other medical factors, about teller and receiver, which are outside their control. Although the diffusion of diseases such as measles has little landscape effect, the spread of other diseases such as malaria, or foot-and-mouth disease in cattle and

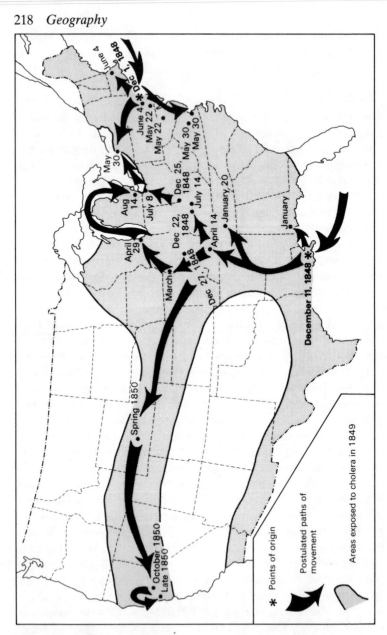

Fig. 14.1   Diffusion of cholera in the outbreak of 1848 in the USA

pigs, or Dutch elm disease, can have profound effects of landscape change.

Two examples of diffusion

The complexity of diffusion processes and particularly of disease is well illustrated by the study of a cholera epidemic in the USA in 1848. Before it became possible to immunise the population against cholera it was a killer disease which spread very quickly via infected water or vegetables washed in infected water. Its spread also depends on low levels of hygiene which allow water supplies to become contaminated. In the USA the disease entered from Europe into both New York and New Orleans within a few days of each other in 1848. From New Orleans the disease was carried up the Mississippi and the Ohio River by the river traffic, first to the larger places in direct contact with New Orleans and then from these larger places it spread to the surrounding countryside. The diffusion pattern was expansionist and hierarchical. Fig. 14.1 shows the spread of the outbreak. The disease eventually reached the Great Lake cities in the summer of 1849 and diffused out to the surrounding areas during that summer. The branch of the disease entering at New York lingered during the cold months in the slums of New York and then spread in the spring of 1849 through the Hudson-Mohawk gap to meet the diffusion branch from New Orleans in the cities of the Great Lakes. The disease also spread south from New York to the coastal cities and from them inland to the smaller towns and rural areas. The major barriers to the westward spread across the Great Plains was the low density of population and the lack of contacts between transmitters and receivers of the disease. This barrier was permeable along the main transport routes to the west and interaction along these wagon routes took the disease through rural areas and ultimately to San Francisco in late 1850. Expansion and relocation diffusion were both important in the passage of the disease through the barrier of the great plains and the Rocky Mountains but in the eastern states with their pattern of cities, towns and villages the diffusion was hierarchical.

At a smaller and more local level, Fig. 14.2 shows a diffusion of quite a different kind. This is the spread of self-service shops through Nottingham, an urban area in Britain, during the 1950s and

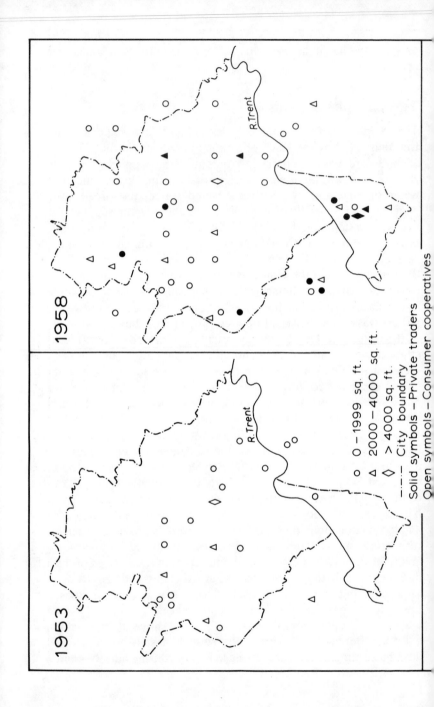

1958

1953

R.Trent

○ 0 - 1999 sq. ft.
△ 2000 - 4000 sq. ft.
◇ > 4000 sq. ft.
—·— City boundary
Solid symbols – Private traders
Open symbols – Consumer cooperatives

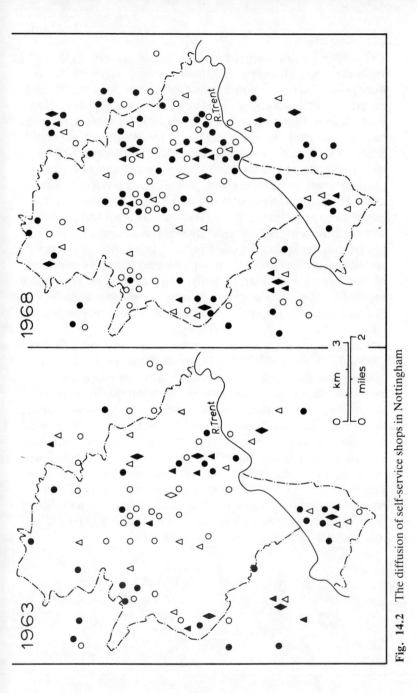

Fig. 14.2   The diffusion of self-service shops in Nottingham

1960s. Four phases occurred in the diffusion process. In the early 1950s there was an innovatory phase when self-service techniques were pioneered by the Nottingham Cooperative Society. The early experiments tended to occur in inner suburban and central city shops. During the second phase multiple retailers, again in the inner suburbs, adopted the new style. It also spread to the Cooperative Society's suburban shops. The third phase is characterised by an increased scale of operation, the beginnings of adoption by independent small shopkeepers and growth of self-service in the suburbs. The fourth phase shows considerable growth in the outer suburbs and a continuing of the trend towards larger scale operation with the closure of small shops in favour of larger ones. There are, in this overall expansion diffusion process, two components: first, a spatial decentralisation of the new techniques from the central parts of the city to suburbia; secondly, the adoption of self-service successively by the Cooperative Society, multiple retailers (for example chain stores such as Tesco) and small independent retailers.

Diffusion processes are more common in explaining social and economic features of the landscape than the physical features. Nonetheless there are instances when diffusion processes can be used to explain vegetational change. For example, in areas newly uncovered from ice sheets, as described in the previous chapter, plants and vegetation become established on till plains. The process of invasion and succession of ecosystems has strong similarities with the process of diffusion. Initial invasion of a plant species is followed by its expansion diffusion to nearby areas but this spread is constrained by various barriers such as adverse soil conditions or prevailing wind direction which may limit diffusion in a particular direction. Diffusion models are used widely to explain how spatial patterns in the landscape change through time.

15

Events

The models of landscape change discussed in the last few chapters have been concerned with progressive changes in the landscape. They are models of gradual or evolutionary change. Landscape change, however, can be sudden and very rapid in certain circumstances. Established processes of gradual change may be disrupted, sometimes violently, by sudden events such as earthquakes, hurricanes or floods, which change the physical features of the landscape and also have a notable influence on the way man uses and creates landscape. A regularly flooded area, for example, may be left to agricultural use although situated within a city. Not all sudden landscape events, however, are the result of 'acts of God'. Some undoubtedly are acts of Man. The policies of government, whether to close a steel plant, build a dam, ban office development in a city, or give grants for hedgerow removal, can produce sudden, unexpected and considerable landscape change, discontinuous with previous processes and interrupting the steady pattern of evolutionary change.

Earthquakes

Shifts in the relative positions of the crustal plates (see Chapter 3) produce earthquakes. There are, and have been for many million years, several hundred earthquakes over the world each year. It is the large ones that result in violent landscape change and fortunately there are usually only one or two of these per year. A movement of tectonic plates generates waves of energy which pass round and through the earth. The earth is made up of a thin, rigid outer skin (several kilometres thick) which is generally thicker under the continents than under the oceans and is termed *crust*. Below the crust is the *mantle*, which is a dense layer of iron and magnesium-rich material about 2900 km deep. Within the mantle is the *core*, with a radius of 3490 km, composed of liquid nickel-iron. An earthquake generates three types of wave called *P*, *S* and *L* waves. *P* waves are compressional waves which spread out from the point (focus) of the earthquake. The point on the earth's surface immediately above the focus is called the *epicentre*. The vibrations in *P* waves follow the line of the wave and are sometimes called 'push' waves. *P* waves pass through the earth, although refracted by the core. Fig. 15.1 shows the paths of *P* waves, and also *S* and *L* waves, through the earth. *S* waves, in contrast, have vibrations at

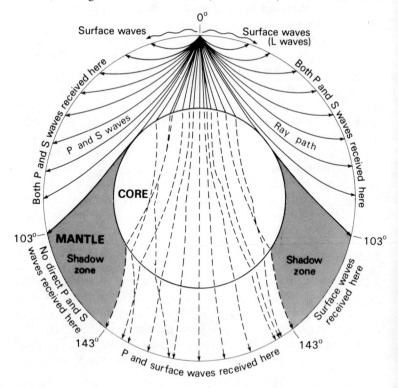

Fig. 15.1 Types and typical paths of waves generated by an earthquake

right angles to the line of movement of the wave and travel more slowly than *P* waves. *S* waves are called transverse waves, or 'shake waves', and do not pass through the core of the earth but pass only through the mantle. The *L* waves, or *Rayleigh* waves, are surface waves limited to the crust of the earth and are the most destructive wave type because of a combined vertical and horizontal movement associated with them.

Earthquakes are commonly measured on two scales. The *Richter scale* is a measure of the energy generated by an earthquake and the *Mercalli scale* is based on the damage resulting from an earthquake. The Richter scale has values from 0 but no upper limit; however no earthquake of intensity greater than 9 has been measured. The scale is logarithmic, so that an earthquake of intensity 6 releases ten times as much energy as one of intensity 5, and a hundred times more than one of intensity 4. An earthquake of class 5 releases about the same energy as the small early atomic bombs. Earthquakes of a magnitude of 7 are rated major earthquakes. The large earthquakes in Alaska in 1964, in Japan in 1952 and in Indonesia in 1977 were recorded at 8.5, 8.6 and 8.9 respectively, with energy releases about 10 000 times more than that of the first atomic bomb. A summary of the twelve classes of the *Mercalli scale* is shown in Table 15.1. This scale relates to the damage done to urban areas, with major structural damage occurring in earthquakes of class 8 and higher. The earthquake of 23rd November 1980 in Southern Italy was of intensity 10 on the Mercalli scale. Fig. 15.2 shows the *isosisms*, lines of equal earthquake intensity, for the event. Over 3000 people were killed and half a million made homeless by the damage to several villages and small towns in the mountains inland from Naples. About eighty-five settlements had more than half their buildings seriously damaged. Not only housing but also industrial and commercial properties were destroyed, and the tourist industry was severely affected. The estimated cost of reconstruction is over £26 billion, but the considerable damage done by the earthquake does provide the possibility for a major investment programme to be undertaken in this extremely poor part of the EEC.

The destructive landscape effects of earthquakes and movements along faults in urban areas are well-known. In rural areas similar displacements may cause streams to change direction or even cause lakes to form by blocking previous river channels. The Alaskan

Table 15.1 The Mercalli intensity scale for earthquakes

I	Not felt except by a very few under especially favourable circumstances.
II	Felt only by a few persons at rest, especially on upper floors of buildings. Delicately suspended objects may swing.
III	Felt quite noticeably indoors, especially on upper floors of buildings, but many people do not recognise it as an earthquake. Standing motor cars may rock slightly. Vibration like passing of truck. Duration estimated.
IV	During the day felt indoors by many, outdoors by few. At night some awakened. Dishes, windows, doors disturbed; walls make cracking sound. Sensation like heavy truck striking building. Standing motor cars rocked noticeably.
V	Felt by nearly everyone, many awakened. Some dishes, windows, etc broken; a few instances of cracked plaster; unstable objects overturned. Disturbances of trees, poles, and other tall objects sometimes noticed. Pendulum clocks may stop.
VI	Felt by all, many frightened and run outdoors. Some heavy furniture moved; a few instances of fallen plaster or damaged chimneys. Damage slight.
VII	Everybody runs outdoors. Damage negligible in buildings of good design and construction; slight to moderate in well-built ordinary structures; considerable in poorly-built or badly designed structures; some chimneys broken. Noticed by persons driving motor cars.
VIII	Damage slight in specially designed structures; considerable in ordinary substantial buildings, with partial collapse; great in poorly-built structures. Panel walls thrown out of frame structures. Fall of chimneys, factory stacks, columns, monuments, walls. Heavy furniture overturned. Sand and mud ejected in small amounts. Changes in well water. Persons driving motor cars disturbed.
IX	Damage considerable in specially designed structures; well-designed structures thrown out of plumb; great in substantial buildings, with partial collapse. Buildings shifted off foundations. Ground cracked conspicuously. Underground pipes broken.
X	Some well-built wooden structures destroyed; most masonry and frame structures destroyed with foundations; ground badly cracked. Rails bent. Landslides considerable from river banks and steep slopes. Shifted sand and mud. Water splashed (slopped) over banks.
XI	Few, if any (masonry) structures remain standing. Bridges destroyed. Broad fissures in ground. Underground pipelines completely out of service. Earth slumps and land slips in soft ground. Rails bent greatly.
XII	Damage total. Practically all works of construction are damaged greatly or destroyed. Waves seen on ground surface. Lines of sight and level are distorted. Objects are thrown upward into the air.

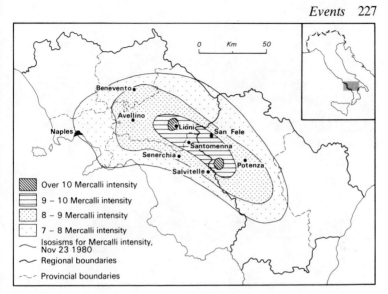

Fig. 15.2 Isosisms for the Southern Italian earthquake of 1980

earthquake of 1964 caused, in some places, a 10 m vertical displacement of the land either side of the fault. In geological history much larger fault displacements, sometimes of 1000 m, occurred. There is debate over whether these displacements took place gradually, so allowing erosional processes to operate during faulting, or whether they took place rapidly enough for erosion and weathering not to mask the result. Certainly they would not have occurred as single events. It is often difficult to decide whether a particular steep slope or mountain edge is the direct result of faulting or whether it is created by erosion along the fault line.

Faults are often classified into normal and reversed faults, as shown in Fig. 15.3. A combination of faults can result in block faulting, creating upstanding blocks bounded by normal faults. Such blocks are termed *horsts*, whilst the intervening basins or troughs are termed *graben*. A *tectonic scarp* is the land form which separates the displaced rock surfaces. These steep slopes or scarps become subjected to relatively rapid erosion and pass through several stages before the line of the fault becomes hidden. Where a series of horsts and graben exist then it is termed *basin and range* topography. A large area of this landscape type exists between the

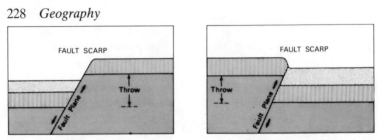

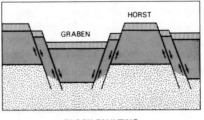

Fig. 15.3 Normal, reversed and block faults

Sierra Nevada in California and the Wasatch mountains in Utah, where there are at least 141 fault enclosed basins. The floors of the basins are filled with gravelly debris eroded from the tectonic scarps. Many of the basins have also been lake filled during an earlier, wetter climate and so often have very flat floors, in stark contrast to the short jagged mountain ranges which exist along basin edges. There are many examples, all over the world, of major landscape features resulting from faulting and tectonic activity, either as simple scarps or as complex features edging great rift valleys or making the edges of mountain chains.

Earthquakes also trigger other landscaping events as well as creating, in their own right, major landscape features and landscape change. One of the most devastating of recorded landslides resulted from an earthquake triggering the collapse of the north peak of Hucascaran Mountain in Peru in May 1970. The ensuing rockslide and avalanche was channelled along a glacial trough and became more fluid as it picked up water and fine gravel moraine material. Within a few minutes the flow of material had moved 15 km, overwhelmed the town of Yungay and killed 25 000 people. Even greater loss of life, probably of almost 250000, occurred with a series of earthquake-triggered landslides in Kansa province, China in 1920. In such cases the potential conditions for a mass movement of material must be present if it is to be triggered by an earthquake.

Rocks and soils of various kinds have different levels of internal cohesion through having different porosity, composition and jointing patterns. Some materials will slide away from a relatively gentle slope, but in other cases stable and steep free faces develop. Civil engineers use the terms *tensile strength* and *shearing strength* when discussing the ability of soil or rock to withstand stresses without slippage. The amount of stress, the slope and the shearing strength of the material are all critical in determining whether a slide will occur. An earthquake, even a small one, may increase the stress above the critical level and so trigger a landslide or rockslide resulting in sudden, often catastrophic, landscape change.

Floods

An existing set of stresses and pressures which is released by some trigger mechanism is a common underlying cause of many landscape events. Earthquakes themselves occur in this way when the stresses created gradually by the movement of tectonic plates are suddenly released. Some floods are caused in this way, but others are due to fluctuations within the hydrological cycle. The classification of types of flood distinguishes flood events which are due to extreme processes in the hydrological cycle from those triggered by other mechanisms such as earthquakes. The classification suggests eight major types of flood:

A Fluctuations within the hydrological cycle
1   Rainstorm – river floods
2   Snowmelt floods
3   Coastal floods due to meteorological conditions

B Other causes
1   Dam or levee failure floods
2   Floods due to the rupture of a glacial lake
3   Floods resulting from landslides and volcanic events

4   Floods induced by land subsidence along coastlines
5   Coastal floods due to sea waves resulting from earthquakes.

In some instances river floods are a normal part of the annual cycle of river activity, with an annual inundation of the flood plain of the river. Such floods may be important in replenishing soil fertility and agriculture may depend on this annual cycle, as shown in Chapter 12 with rice cultivation. The early agriculturalists, 8000 years ago in the valleys of the Tigris, Euphrates, Nile and Indus, relied on regular floods to provide irrigation. Their ability to manage the flood allowed these early civilisations to flourish. Such regular and reliable floods are perhaps better considered as part of a cycle rather than as an isolated event, but individual floods may be considered as an unpredictable event.

A rainstorm-river flood may be defined as a discharge which exceeds the channel capacity of a river and then proceeds to inundate the adjacent floodplain. After the rain the amount of water which runs off into the river is the critical factor determining whether or not a flood occurs. The amount of runoff depends on:

1   Amount of rainfall
2   Steepness of slope
3   Character of soil and subsoil, including moisture content
4   Presence or absence of vegetation
5   Degree of man's modification of the river cycle.

We have already seen in Chapter 10 that long steep slopes erode more than short shallow ones and that runoff is similarly greater on steep slopes. The character of the soil governs the extent to which rainfall is absorbed, or infiltrates, into the soil. Soils with large spaces between individual particles, such as gravels, are permeable allowing maximum infiltration. The amount of infiltration is termed *infiltration capacity*. Clay soils, unless heavily cracked, allow relatively little infiltration. The underlying subsoil is important as it may be of different permeability from the topsoil. If an impermeable subsoil underlies a permeable topsoil then water may flow down-slope at the subsoil junction, creating *subsurface storm flow*. The infiltration capacity of the soil changes during a rainstorm and decreases with soils swelling as they get wet. Vegetation is also important in intercepting rainfall and reducing runoff. In heavily vegetated landscapes such as tropical rain forests, surface runoff as a proportion of rainfall is small, but on thin vegetation such as on desert margins around the Sahara (see Chapter 8) runoff is high when occasional rain falls. The clearing of vegetation by man, whether for subsequent agricultural use of the land or whether for urban development, increases runoff considerably. With greater runoff there is a greater chance of floods taking place. The construction of roads, pavements and buildings in a stream or river collecting area increases the chances of the stream or river flooding. A study of the American city of Jackson on the Mississippi showed that where urban land uses had covered stream water catchment, the chance of flooding was 4.5 times greater than in comparable stream catchments with agricultural land uses in surrounding rural areas. Man can be a significant factor in increasing flood frequency.

The likelihood of a river generating a flood also depends on the stream and river network (see Chapter 2). With river networks that are highly branched the flood is higher but of shorter duration than

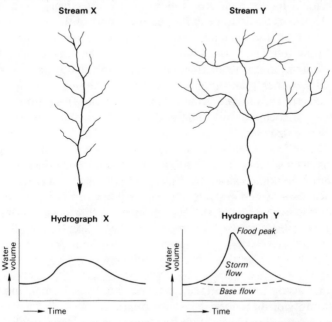

Fig. 15.4 Contrasting flood hydrographs associated with different river networks

less branched networks. Fig. 15.4 shows two networks and the graph or *hydrograph* of water volume (discharge) through time. In the less branched network, stream X, the water surges arrive successively in the main stream and so the flood on the main river lasts longer.

The hydrograph shows the amount of water passing a point on the river. After heavy rain there is a time lag before the level of the river rises to the flood peak. The rate of decline of discharge is much slower than the rate of increase but eventually the river level returns to normal, which is termed base flow. The length of lag time between rainfall and the flood peak passing depends on the size of the river basin and may vary from an hour or so on small first order streams to several days on large higher order rivers such as the Danube or Mississippi. This means that often several days warning may be given of floods in the lower Mississippi or on other large rivers and defensive action may be taken (see Chapter 17). The movement of the flood peak also depends on the movement of the storm. If for example storm movement is effectively up the river valley then the flood peak in the lowest part of the valley may be earlier than if storm movement was effectively downstream.

Snowmelt and coastal floods may also be predictable. With snowmelt floods the main variable is the speed at which snowmelt occurs, rather than the quantity of snow melting, and snowmelt speed is not always predictable. Coastal flooding from storm surges, usually associated with high tides, can be forecast and attempts can be made to lessen the effects before the flood arrives.

Other flood events are far less predictable and rather less common. Dam bursts may be triggered by earth movements but are more commonly caused by faulty design. Glacial lake drainage is more complicated and not fully understood. In some cases the ice melts in summer and the lake empties, filling again during the spring. There are several types of glacial lake, each of which provides its own flood hazard. The five most common types are:

1　Lakes on the ice surface; such lakes are usually small and their emptying does not produce major floods
2　Lakes within the ice surface; again these tend to be small
3　Lakes in tributary ice-free valleys dammed by a glacier or ice

sheet; these may be large features, much larger than the two earlier types
4   Lakes dammed by a glacier against the valley wall
5   Lakes created by ice avalanches across a valley; these may be relatively temporary lakes which empty suddenly.

The emptying of these lakes may have considerable effect on the physical landscape but such lakes are often in areas with little, if any, population.

The most unpredictable flood, and the one which often causes most damage, is coastal flooding from a sea wave (*tsunami*) generated by a submarine earthquake. Tsunami are relatively shallow, very long wavelength waves in the open oceans but their height increases, sometimes to 20 m, as they reach shallower water. The giant volcanic eruption and associated earthquakes of the Krakatoa explosion in 1883 generated a tsunami about 30 m high when it hit the coasts of nearby islands. Because coastal areas are usually heavily-used and densely-populated, and because of the unpredictability of tsunami, loss of life and destruction of property is considerable from large events of this type. Coastal flooding by seawater causes damage to agricultural and urban lands. The cost of flood damage in California, associated with tsunami after the 1964 Alaskan earthquake, was estimated at $10 million at 1964 prices.

Government decisions

Landscape change as a result of government decisions can be abrupt, as the new policies effectively create a disturbance in existing processes. The decision of the governments of the Oil Producing and Exporting Countries to raise prices massively in 1973 is typical of such action. The policy decision by OPEC to reorganise its price structure changed many of the processes in economic geography and introduced new elements of landscape change. Many countries have changed their energy-use pattern: coalmines which would have closed have remained operational; oil exploration in non OPEC member countries has increased; nuclear energy generation programmes have been increased; people's travel habits have changed; recreational trips in private cars are fewer and shorter; industrial production costs have risen steeply – the list of changes could go on for many pages. The changes all suggest a major change in direction of economic processes as a result of the one initial decision.

An example of a more local but again far-reaching decision was that of the United States Government, through the NASA organisation, to locate a manned spacecraft control centre in Houston, Texas. Within a few years in the early 1960s over 200 000 new residents were attracted to the city to work in the centre itself and in more than a hundred aerospace firms which established new offices and factories in Houston. Local income and property taxes were boosted and extremely rapid growth spread through the whole local economy, from residential building to services and shops for the new population. Other sites considered for the centre were Boston, Jacksonville, Tampa, and St Louis, besides other sites in Texas, California and Louisiana. The growth of the city of Houston in the last twenty years would have been very different if one of the other cities had been chosen for the NASA centre. There were several reasons for the choice of Houston over the other places, including the gift of the land to the government; the presence of existing research centres and universities in the area; port facilities; the existing large labour force; and the fact that Texans held key government posts of Vice President and Speaker of the House of Representatives at that time. Debate continues over the relative importance of the various factors, including the last one, in the final decision. A particular decision or a few major events are often decisive in the process of city growth and the evolution of the city landscape. Such events may result from some form of government decision which can stimulate either growth, as in Houston, USA, or decline, as in Consett, Co Durham in England, where the major employer, the state-owned steelworks, closed. Some types of government decision may be just as cataclysmic to the urban and rural landscape as earthquakes or tornadoes.

Events, cycles and stages

The event, cycle, stage and diffusion processes discussed in this and the previous three chapters are not unrelated to each other despite their division into separate chapters. It can be useful, to distinguish amongst them as has been done here, but the division is arguably artificial. Events might be considered as the start or finish of a particular stage of development, whilst the stages themselves may provide feedback, so producing a cycle effect. As with all geographical processes, scale is important. Processes of large scale landscape change may show a staged development; more detailed processes making up the smaller scale changes may be of a different type, perhaps a cycle. Despite the difficulties in their use, models such as those described in earlier chapters provide a framework against which changes in the real world may be compared. In this way they help us to explain why landscape change takes place.

16

Man's Ability to Initiate Landscape Change

Man has a strong influence on landscape change and is often directly responsible for it. In some countries a framework of land-use planning legislation within which to direct and monitor landscape change has developed. Although this type of legislation has only become widely accepted in the last thirty years, and the environmentalists who are concerned with man's abuse of landscape have operated effectively only for twenty years, there is a long history of man influencing and changing the landscape. As soon as man began to use and exploit natural resources he had an influence on the process of landscape change. The importance of man as an active agent for change was stressed in the early chapters of this book and has been implicit in many later chapters. The purpose of this chapter is to enlarge upon some of the ways that man is currently changing the landscape, and more importantly creating socially unacceptable change through pollution of water and the atmosphere.

Technology's impact on the environment

Each new technology invented, adopted and diffused through society has had an impact on man's environment. In traditional societies man's ability to alter the physical environment was limited. Small scale agriculture changed vegetation patterns slightly and water management methods only had a small effect on river regimes. A lack of technological expertise limited man's effectiveness in overcoming environmental difficulties and changing the landscape to his advantage. Such societies were, and in some cases still are, limited in their ability to manipulate the landscape and therefore less likely to disrupt the balance in ecological systems than industrial and technologically advanced societies. Waste from these traditional societies, for example, is largely organic and is relatively easily decomposed and recycled by environmental processes. Durable features in the landscape, such as buildings, last a long time and even when the buildings are destroyed the materials are often used again in other buildings. Even in traditional societies the landscape is changed over the years, but it is altered relatively slowly and environmental processes change to accommodate man's influences. It would be wrong to assume that traditional societies, even with low population densities, have no influence on the environment and landscape.

In many of the tropical grassland regions of the world, traditional societies, through the planned use of fire, have changed the environment in many ways. The firing of grassland in the dry season provided fresh growth for grazing and helped limit the invasion of bushes and trees (see Chapter 7). Even in tropical forests, where burning is more difficult, deliberate tree-felling and burning creates clearings in the forest which are used for agriculture. The fertility of such forest plots declines rapidly due to the heavy rain once the trees are removed, and it is necessary to repeat the clearance exercise elsewhere after a few years. It is possible after many years to return to earlier used areas and clear them again and use the plot a second time. This *slash and burn* or *swidden* type agriculture, when carried on for many centuries as it has been in tropical Africa, has reduced the area of 'natural' forest and allowed secondary forest, typified by different tree species, to become dominant over large areas. As a general rule, in both grasslands and forests man tends to reduce the number of natural plant species with a low tolerance of moisture or temperature fluctuations and to increase the number of vigorous plant species with a high tolerance of environmental fluctuation.

In technologically more advanced societies the capacity for landscape change is considerable and change can take place rapidly. Although a far cry from the agricultural technology of the late twentieth century, the farm society that settled in the centre of America in the nineteenth century had technology a quantum leap more sophisticated than the traditional Indian societies. Vast areas of the United States were deforested very rapidly. Fig. 16.1 maps the tree cover of a 10 km by 10 km area of South West Wisconsin at four dates between 1831 and 1950. Before the new technology appeared the area was almost totally covered by deciduous hardwood forest, except for a small area of grassland in the south west corner. By 1882 settlement had occurred and agriculturalists had cleared for cultivation about seventy per cent of the forest land. In the next twenty years the forest had been reduced to about sixty plots of trees comprising less than ten per cent of the total. By 1950 this had been reduced to about five per cent, but this decrease took place not with the removal of complete woodland areas but through

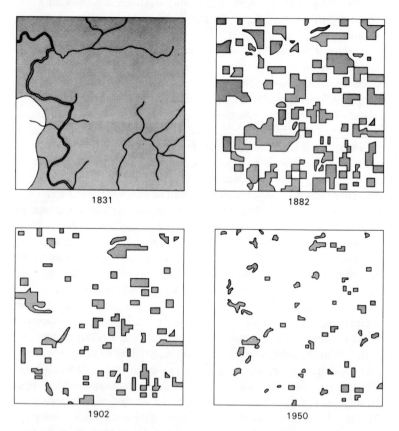

1831   1882

1902   1950

Fig. 16.1   Reduction and fragmentation of woodland area in a small part of Wisconsin, USA

the reduction in size of individual woods. Over 120 years the landscape of this small area of the USA has been transformed totally by man's agricultural activity.

The reduction in woodland has brought in its train other changes to the environment: a change in the soil water balance has resulted in many streams drying up; the species composition of the woodland changed as grazing animals were fenced out and limited to the cleared land; the animal species composition of the area changed, with an increase in species adapted to living on the edge of woodlands rather than within woodland habitats.

Only as man has learnt more about ecosystems and their internal relationships has it proved possible to disentangle the complex relationships in environmental change. A relatively small and apparently insignificant change at one point may result in a critical change elsewhere which alters some fundamental balance in the ecosystem. In modern industrial societies with an advanced technology man's control over the environment has increased and some ecologists, who think only in terms of physical environments, suggest that by building cities man is insulating himself from the environment. Building cities undoubtedly changes the environment, but a new one is created just as stimulating as the one it replaced. Problems can arise in the connections between the new man-made environments which change quickly and the older man-influenced environments where change is often slower. In modern agriculture, technological advances have allowed improved plant growth, control of pests, redirection of water movements and even the reshaping of the land. In some of these activities outputs from the system become the unwanted inputs into other systems, creating pollution and similar disruptive environmental problems.

Industries using modern technology consume large amounts of natural resources, and in transforming these resources into goods useful to society, a considerable amount of waste product results. In areas where technologically based industries are concentrated, such as large cities, waste is produced faster than it can be absorbed by environmental processes and the water and air around us becomes polluted. In non-nuclear energy production particularly, large amounts of waste are created compared with the usable energy created. Waste gases are released into the atmosphere from power stations and car exhausts, liquids are released into the hydrosphere and solid waste may result in spoil heaps and large dumps. Technology is now advancing to a stage when these waste products cease to be considered as useless but may be used, or *recycled*, to produce new goods. The waste heat from energy production may be reused elsewhere rather than being released into the atmosphere and hydrosphere. Metallic waste can be sorted and reused; paper waste from cities can be recycled to produce usable paper again; organic wastes from cities can be processed and used as agricultural fertilisers. Technology is now coming to grips with this recycling, which can be viewed as either system feedback, as in paper use, or as controlling the interrelationships between systems, as in the case of processing organic waste for fertiliser. It may be some decades before technology can solve certain problems, however, and in the meanwhile *pollution* occurs. Some environmentalists would therefore have us halt economic progress and return to more traditional societies. Many geographers would advocate the more positive solutions associated with improved recycling technology.

If pollution is here to stay for some decades it is worthwhile considering how pollution occurs and how it may be controlled in the short term. It is also worthwhile distinguishing between the material that creates pollution and the processes which move and distribute this material.

Water pollution

Water pollution problems in general result from land-based activities which dump waste into a river or other water body. Fig. 16.2 is a diagram often used to show how the quality of water in a river may be affected by many quite different activities. Even natural processes of erosion in mountain areas (1) release minerals which may be harmful to water users, or may simply reduce the quality of the water. Similarly dead vegetation (2) may be a source of water impurities. Much more important, however, are the pollutants which enter the river from urban sources. Most sewage works (3) do not remove all water impurities; street drains (4) empty water from city streets directly into the river, certainly after storm events and in some cities all the time. Industrial waste water (5) adds further pollutants to the river. More specific sources of pollution may be power stations (7) returning warm water to the river; strip-mining

Fig. 16.2 Land uses having an adverse influence on water quality

activities (6) (see Chapter 3) releasing mineral pollutants; oil spills
from river transport (9); waste tipping (10) affecting ground water
storage and quality; and industrial gases washed out of the air by
rain (8). There are also sources of pollution in agricultural land uses
arising from the spraying of crops with pesticides (11); the addition
of fertiliser (12); and the general runoff of farm waste (13). Various
laws may be introduced to limit the types and amounts of potential
pollutants which may be released into rivers and water bodies, but
pollution still occurs from all the sources shown in Fig. 16.2.
Landscapes with different patterns of land use will have very
different potential for water pollution.

Organic pollutants in water are acted on by organisms called decomposers which use oxygen in the water to remove them. A measure of the amount of organic pollution is the amount of oxygen in the water which is needed by the decomposers in order to remove the organic material. This is called the *biochemical oxygen demand* (BOD). BOD is a measure of the contamination of water, and in extreme cases levels can reach 30 000 parts per million, which means that enormous quantities of oxygen are required by the decomposers to use up the organic pollutants.

When polluted water containing high levels of BOD enters a river the organic pollutant is gradually removed by decomposer organisms in the river and a characteristic sequence exists downstream from the pollutant source. At the point of discharge decomposer bacteria begin to consume the organic material and increase in numbers with the plentiful supply of food. Oxygen is used up rapidly and faster than it is replaced by the activity of aquatic plants and transfer from the atmosphere. As the pollutant is moved downstream so it is reduced in quantity and the oxygen demands of the decomposers are reduced. At a critical point the oxygen replacement rate exceeds the removal rate and then the river becomes restored to its original condition. The sequence of rapid oxygen depletion followed by gradual replacement is often called the *oxygen sag curve*. The length of river required to absorb a given quantity of pollutant (the length of the oxygen sag curve) is important in the planning of waste disposal, since the addition of more organic material before the river oxygen level returns to normal can be disastrous and *anaerobic* conditions may result. Decomposer bacteria can be of two types, *aerobic* and *anaerobic*. Aerobic decomposers consume oxygen but are effectively cleaners of the water. If oxygen levels fall too low, then aerobic bacteria cannot operate and *anaerobic* decomposers take over. These bacteria do not operate in the same way but generate ammonia, methane and hydrogen sulphide, thus turning the water course into a rotting putrid mess from which gases bubble and in which few plants and animals can live.

Organic pollutants are not the only ones causing reductions in water quality. Various suspended solids, nitrogen in various forms, and other chemicals all may find their way into rivers and water bodies. Table 16.1 shows the levels of pollution associated with

Table 16.1 Range of concentrations of selected substances in urban
stormwater and water runoff from agricultural land

A Urban storm water

Pollutant	Concentrations in parts per million
Biochemical oxygen demand	1.0–700
Suspended solids	2.0–11 300
Organic – nitrogen	0.1–16
Ammonia – nitrogen	0.1–2.5
Total phosphorus	0.03–42
Chloride	2.0–25 000
Oils	0–110
Lead	0–1.0

B Agricultural land runoff

Pollutant	Pasture	Arable	Woodland
	Concentrations in parts per million		
Biochemical oxygen demand	6–17	4–31	4–7
Suspended solids	11.8–840	286–4200	45–132
Total nitrogen	2.5–8.5	15.0–37.0	2.4–5.1
Total phosphorus	0.24–0.66	0.18–1.62	0.01–0.86

rainwater running off urban streets, together with some less detailed values for water draining off three rural land uses. Mention
has already been made in earlier chapters of the removal of surface
material which takes place on cultivated land. This soil erosion is
shown in Table 16.1 as the high values for suspended solids in water
from cultivated land. The chemical element, and particularly nitrogen content, of this runoff may also be high due to additions of
fertiliser to the land. The very high chloride levels shown in Table
16.1 are caused by urban runoff taking place when various chemicals
including salt have been used to clear snow and ice from the streets.

Atmospheric pollution

The atmosphere contains pollutants just as hazardous as those in
water. The atmospheric pollutants which are harmful to people are

carbon monoxide, various sulphur oxides, hydrocarbons, nitrogen oxides and dust. Most of the carbon monoxide and hydrocarbons come from transport activities and emissions from oil and petrol engines. The sulphur oxides and nitrogen oxides also come from fuel combustion, and frequently from solid fuels used in factories and power stations. Transport and fuel combustion at industrial premises account for about two-thirds of the pollutants in the atmosphere of Western Europe or North America. These pollutants are normally present in very small quantities, but meteorological processes can result in their concentration and create health hazards.

Usually, warm air from factories and engine emissions rises and as it rises it cools and expands. Other air falls to take its place, mixing occurs in the atmosphere and harmful pollutants are dispersed. In this usual situation the ground and lower part of the atmosphere are warmer than the upper parts, thus encouraging the rising air and dispersion of harmful substances. Sometimes, however, a *temperature inversion* occurs, when the ground level layer of air is cooler than that above it, so inhibiting the ground air from rising. Fig. 16.3 shows the change in temperature with height for the normal situation (dotted line), in which temperature falls by ap-

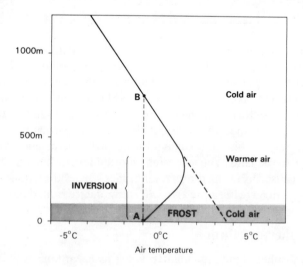

Fig. 16.3 A simple temperature inversion

proximately 0.6°C per 100 m, and the case of the inversion, when cold air is trapped at ground level. An inversion can occur during the night if the sky is clear and the air calm. The ground surface loses longwave energy into the atmosphere above it, ground surface temperatures drop and the surface air layer becomes colder. In Fig. 16.3 the ground temperature has fallen below freezing point. From ground level to about 300 m the temperature increases with height, and then above this level temperature declines with height in the usual fashion.

When this type of inversion develops over an urban or industrial region from which air pollutants are being passed to the atmosphere, then the polluted air stagnates and a heavy fog or smog can be created. The top of the inversion, the *lid*, marks the level at which the pollutants are held and below this the air pollution builds up. Over a city this build-up of pollution may be rather like a dome, with increased intensities of pollution at the centre. Even when inversions occur infrequently there may still be a build-up of air pollution over a city simply because there is a large output of pollutant materials. When a steady wind spreads this pollution into a *pollution plume*, then places downwind of the city may suffer air pollution. Just as smoke from a domestic or factory chimney often plumes downwind, so pollution from a whole city may be similarly spread. With such plumes the heavier dusts may come down as fall-out many kilometres from the pollution source.

Urban air pollution can reduce visibility and also act as a screen to incoming radiation. In the winter months in British cities, air pollution may result in the loss of fifty per cent of incoming solar radiation and frequently in winter, losses of over twenty per cent occur. Urban areas also have other effects on local climates and some environmentalists extend the idea of atmospheric pollution to include these effects. The build-up of heat in cities can create temperatures in the city centre 3°C or more higher than in the surrounding outer suburbs. This temperature differential is often greatest at night. During the day the concrete in pavements and buildings in the city centre receives and stores heat. As air temperature falls this heat is radiated by the buildings and causes a slower lowering of temperature in the city centre than in the suburbs. Other features of urban climate are higher precipitation due to the presence of more dust to act as condensation nuclei around which precipitation may form; lower windspeeds due to the presence of large buildings; and, of course, increased cloud cover and fog. Usually, the larger the city the greater is its effect on climate.

Air and water pollution are inseparable. Rain occurring in areas of high air pollution itself becomes polluted and may fall as *acid rain*. The acidity and alkalinity of liquid is measured on the *ph* scale, which runs from 7 to 0 on an acidic scale and from 7 to 14 in alkalinity. Acids have low values, alkalis high values and the neutral value is 7. The scale is logarithmic so that a difference in 1 in *ph* values means a difference of ten times in acidity or alkalinity. Lemon juice has a *ph* of about 2.5 and wine of about 4. Rainfall is normally slightly acidic since some atmospheric carbon dioxide is dissolved, but when the air is polluted with sulphur oxides and nitrogen oxides the rain becomes more acidic. Values for *ph* below 4 now occur in the industrial north-east of the USA and in industrial Europe, whereas twenty years ago values of 5 were considered exceptionally low. Since the mid 1950s rainfall has become ten times more acidic in many urban industrial regions. Values as low as 2.5 have been recorded for particular rainstorms. One of the many problems of acid rain is that it does not always fall in the country responsible for the pollution. Pollution plumes may spread many hundred kilometres across national boundaries and so generate acid rain a long way distant from the pollution source. Acid rain can cause water pollution problems, killing the natural plant and fish populations, and it can also accelerate rates of deterioration of building materials, particularly limestones and marbles. Unless the amount of nitrogen oxides and sulphur oxides entering the atmosphere is more strictly limited in industrial areas, acid rain will become increasingly common and will increase in strength, with resultant disastrous effects on many species of wildlife.

Pollution in an ecosystem

The movement of material through ecosystems means that if pollutants enter an ecosystem cycle and then pass through it, they may build up toxic concentrations at specific points in the system. DDT is an example of a pesticide, or toxic compound, which can build up in food chains in ecosystems. Thus very low concentrations (perhaps .0005 ppm) may be present in the water due to runoff from areas where DDT has been applied. This concentration can build up in the bodies' tissues through plankton (0.04 ppm), snails (0.26 ppm), clams (0.42 ppm), fish (2.0 ppm), herring gulls (5.0 ppm), to large birds such as cormorants with very high concentrations (25.0 ppm). At successive levels in the food chain of a particular habitat the concentration of DDT in body tissue increases. DDT takes a long time to be broken down in the environment; the amount is roughly halved each fifteen years, assuming no new inputs occur. It is estimated that in 1970 there were about 1.5 million tonnes of DDT then on land and 0.5 million in the oceans. Annual production was 100000 tonnes per year, of which about 68000 tonnes entered the ground and the remaining 32000 tonnes remained in the atmosphere after spraying. Of this 32000 tonnes about one third returned to the land via rainfall and two thirds were transferred to the oceans, again through rainfall. There was also a small transfer from land to oceans through stream and river runoff. Although these movements are sometimes called the DDT cycle it is not a true cycle comparable to those discussed in Chapter 12. DDT enters the environment from the factory and gradually builds up in the oceans as there is no link back to the land from the oceans. Eventually DDT breaks down into harmless components, but this is a slow process and the DDT can remain active whilst it passes through the food chain and builds up in large birds and animals. Society increasingly seeks to control the use of chemicals such as DDT and so control some aspects of landscape change.

Over the past twenty years geographers have generally become more active in the study of environmental pollution and the links that occur amongst different land uses and landscapes. Only as geographers and other scientists have begun to understand the processes operating in landscapes has it become possible to trace the passage of particular materials through the environment and to see precisely how activities and landscape change in one place have an influence on what happens elsewhere.

Man's Ability to Control Landscape Change

The examples of the last chapter showed the apparent ease with which man can influence and disrupt environmental processes and cause landscape change. As more becomes known about man's role in landscape change, so it is realised that individual actions may have wide-ranging effects on the landscape. Such actions may create costs which have to be paid by the whole community and so, not unexpectedly, attempts are made to control those activities which have a potentially disruptive effect. Governments introduce policies to reduce the input of pollutants into streams or the atmosphere; to reduce flood hazards; to control the use of chemicals such as DDT; and they also design land-use planning policies to control landscape changes. Through such governmental intervention it is hoped that the environment will not be abused by small groups but be conserved for everyone.

Land-use planning policies

Land-use policies have a second purpose: to improve living standards by creating a pattern of land use which allows everyone to have access to the services, such as health and education, which they need. Land-use planning is an important process responsible for landscape change in the late twentieth century. Urban landscapes in Britain, for example, are almost totally the result of planned activities by various government agencies. Rural landscapes are also controlled, but in a less formal way.

The gradual acceptance of land-use planning and its steady

growth in influence over the last eighty years is particularly well illustrated by its development in England and Wales. In the late nineteenth century a pressure group emerged who advocated new model settlements, called garden cities, in which land uses were controlled. They also called for better quality housing in existing cities and new, less dense planned layouts for suburbs. At the same time there was increasing concern over the health and housing conditions in the British industrial cities which had grown rapidly in the middle decades of the nineteenth century. Town planning first developed as a response to these health and housing problems on one hand and to pressures from a small group of social reformers on the other.

The first real piece of urban land-use planning legislation was The Housing, Town Planning, etc. Act of 1909. This piece of government legislation gave local district authorities limited powers to control housing standards in new residential developments. A second Act in 1919 extended the earlier legislation by making it compulsory for larger authorities to prepare plans for new developments. Later Acts of 1929 and 1932 further extended the planning powers of local government and allowed joint planning activity between authorities. Actual progress in preparing plans, however, was very slow and despite the various possibilities under the Acts, by 1942 only about five per cent of England was affected by town planning schemes.

This changed radically after the 1947 Town and Country Planning Act, which made the preparation of plans showing present and future development compulsory for county authorities. New developments needed the permission of the local planning authority, who could accept or refuse these applications in the light of the overall plan for the country. Although there have been some modifications to this legislation in the past thirty-five years, all of Britain is still subject to development control, and permission is required for new urban and industrial development and substantial changes of land use.

Alongside the 1947 Town and Country Planning Act was the New Towns Act of 1946. This act allowed the building of New Towns which, ideally, were to be new self-contained communities set down in the countryside where housing, jobs and all necessary services would be provided. They represent a reaction against the overcrowded conditions in many cities; the pressing post-war housing and redevelopment problems faced by these cities; and the pattern of long journeys to work resulting from concentrating urban growth in existing large cities. There are now more than thirty New Towns in Britain. The earliest, designated in the late 1940s, were designed to take the overspill population from London and Glasgow. Among the towns created were Harlow, Crawley and Basildon around London, and East Kilbride south of Glasgow. Later new towns were designed to stimulate the general economic revival of depressed regions by providing industrial sites which would attract new firms and new housing for the workers in these firms, for example at Warrington in England, Newtown in Wales, and Livingstone in Scotland. Around two million people now live in British New Towns and it is difficult to imagine the British urban landscape without the influence of the new architectural and planning ideas piloted in these New Towns. The New Town idea was widely adopted in other countries once the success of the British scheme was appreciated. Although the various governmental provisions differ from nation to nation, the creation of new towns has become a part of the land-use planning approach of countries as different as Brazil, Sweden, Israel and the USSR.

Another planning idea originating in Britain and subsequently widely adopted elsewhere is that of Green Belts around cities to constrain and contain their growth. Green Belts have also been used to prevent adjacent towns merging and in some cases to conserve the special character of historic cities. A Green Belt is designated as a band of land around the outer suburbs of a city which cannot be developed. Such belts clearly influence the processes of urban growth. Suburban expansion, as suggested in the models discussed in Chapter 10, is no longer possible. Although Green Belts are successful in containing urban sprawl, they result in growth pressures being pushed beyond the suburbs and its Green Belt into villages within commuting distance of the city. About ten per cent of England and Wales is currently under Green Belt control. The London Green Belt was suggested in the 1944 plan for Greater London drawn up by Sir Patrick Abercrombie. He envisaged a belt about 8 km wide, providing a barrier to growth and a recreational area for Londoners. The Green Belt which was finally designated was rather wider than originally envisaged and had the added function of protecting agricultural land. From the late 1950s onward most British cities defined Green Belts of protected land around their suburbs, simply to stop further urban development. The idea of Green Belts, like that of New Towns, is now widely used throughout the world as a method of land-use planning.

Within Britain there are also several other types of special category land which may be designated by local or national planning agencies. The ten National Parks contain extensive tracts of open countryside which have been defined to preserve their natural beauty and to encourage informal open-air recreation. Again the idea has been taken up in many countries including Japan, Thailand and the United States. Including National Parks, Green Belts and other special areas such as Areas of Outstanding Natural Beauty, Nature Reserves, Forest Parks and Areas of Great Landscape Value, over forty per cent of Britain's land surface falls into areas which are subject to stricter control than is given by the general Town and Country Planning legislation. The areas under these special controls are shown in Fig. 17.1. Many changes in the landscape, therefore, are very strictly controlled with severe limitations on development and land-use change.

This type of land-use planning defines where development cannot take place and it influences landscape change by the refusal of permission for land-use change. An alternative approach adopted in some countries is to define areas where development may take place and plan the changes in those areas. The overall land-use plan for the city of Copenhagen, Denmark, defines a series of five corridors or fingers stretching out from the city centre. Along these fingers development is encouraged and a series of nodes of planned development have been defined in each finger. The development fingers are based on transport routes radiating from the city centre with a maximum journey time of forty-five minutes from the tip of the finger to the city centre. Between the fingers of development are green wedges of farmland and recreational open space from which urban development is excluded. As the plan has operated and been extended, so planned development along the corridors has included industry and office land uses, and not everyone now journeys into the city centre to work. The plan has encouraged the decentralisation of many city centre functions to suburban areas and this has been further encouraged by the building of an outer ring motorway

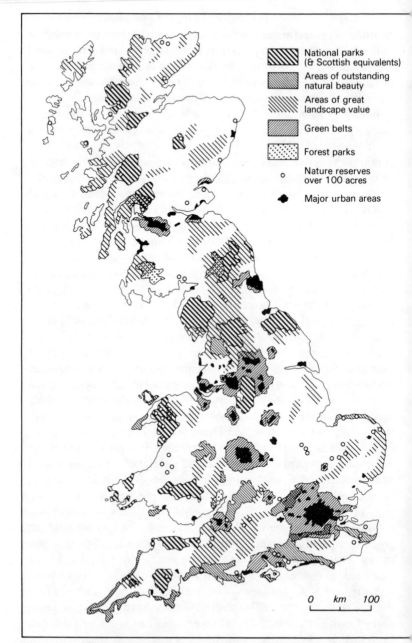

Fig. 17.1 Areas of Great Britain subject to development controls addition to those provided for in general land-use planning legislation

which has joined together the outer parts of the growth corridors, allowing greater communication between the fingers of urban growth. Almost all major cities now have consolidated land-use plans which direct the future pattern of city development and effectively specify what the future urban landscape will look like.

Flood control policies

Alongside the comprehensive land-use planning mechanisms there are, in most countries, specific planning agencies whose responsibilities are limited to controlling specified environmental processes. The control of floods typically is the responsibility of a specific agency. Several possible methods exist for reducing flood damage and protecting people and property. Approaches to the problem can be divided broadly into two groups. First there are *structural schemes*, which include a whole range of engineering methods such as altering river channels to make them less likely to flood and altering buildings to make them more able to withstand floods. Secondly, there are *behavioural schemes*, which include methods such as paying for flood damage out of public funds; providing better insurance cover; setting up a flood warning scheme; or influencing land-use plans to halt building in areas with a high risk of flood. In most countries, and particularly in the USA where many of the cities on major rivers are prone to flooding, there has been a change in approach over the last thirty years with a movement away from structural schemes towards the increased use of behavioural schemes.

The reasons for this change in emphasis of flood protection agencies are threefold. The failure of old structural adjustments to control flooding is certainly a major factor. In several major floods previously engineered flood protection schemes simply failed. With the loss of confidence in such schemes it was natural for agencies to explore other approaches to flood protection. Secondly, new large scale structural flood control schemes of a size and complexity to be effective have become too expensive to be considered, except in a very few cases. In contrast many of the behavioural schemes are low cost, certainly in comparison to structural schemes. A third reason is that improved knowledge of the processes responsible for flooding has increased the effectiveness of behavioural schemes such as flood prediction and warning. Traditionally, flood control has been very much the responsibility of the civil engineer, but increasingly the geographer has become involved in designing control schemes.

In any type of scheme it is impractical to attempt complete flood control. Generally structural approaches change the frequency and magnitude of floods, so reducing damage, and engineering schemes are designed to control floods up to a certain size. Damage still results when floods larger than this critical design size occur. Four major types of engineering schemes exist:

1 The building of embankments and levees
2 Channel enlargements
3 New channel cuttings such as flood relief channels and cutoff channels
4 Flood storage reservoirs

Flood embankment and levee systems are designed to restrict flood water within the river channel. These features range in scale from small local embankments to the major levee building scheme on the Mississippi where constructed levees run for over 4500 km. These schemes are widely used and can be very effective, although they tend to instil a false sense of security in residents living on the 'safe' side of the levee. Floods still occur which are larger than the levee controls were designed to withstand, and they can be very serious.

Channel enlargement schemes are also designed to confine flood water, but the river channel has to be dredged regularly to clear sediment brought back by river processes. A second danger is that whilst a specific locality may be adequately protected, places downstream may suffer worse floods than previously.

Flood relief channels, such as those on the river Welland near Spalding in Britain, involve the digging of a new river channel to take flood water. One benefit of such an approach is its relative cheapness. A related approach is the digging of new channels to intercept the main river and divert it down a new channel, as on the Great Ouse at Ely, also in the Fenlands of East Anglia.

The fourth approach is to build flood storage reservoirs which hold back flood waters, releasing them gradually after the flood danger has passed. This approach has a long history of application, dating back to Roman times. In the sixteenth century Leonardo da

Vinci suggested the idea to protect Florence from flood. In the last hundred years such schemes have been used in many parts of the world. The advantage of the scheme is that it does not simply push the flood problem on to the next community downstream, but there are two main difficulties. First, such schemes tend to be large and expensive. It is sometimes necessary to flood large areas upstream of the dam in order to protect downstream land. In the USA it is suggested that two million hectares of land have been flooded in reservoirs in order to protect five million hectares from flooding. The second problem is that when sediment accumulates in a reservoir, the storage capacity is reduced and its effectiveness as a flood control mechanism is lessened.

A structural but non-engineering approach to flood control is the *flood proofing* of buildings. Lower floors of buildings may be used as car parks, pumps built into basements, or buildings raised above flood level. Widespread flood proofing is only possible slowly and as old buildings are replaced. This approach has been used to protect commercial property, particularly shopping centres, in a number of Mississippi Valley cities, but it is not a widespread approach to flood damage prevention.

All these structural approaches have a very clear effect on landscape. Not only do they influence the landscape during times of flood by controlling the flood, but their presence is very obvious at other times. For example, different land-use and occupancy patterns develop on either side of levees. The building of these schemes involves planning activity by some sort of flood protection agency, which is usually government sponsored and financed.

Behavioural schemes rely on the personal or group planning activities of individuals taking action prior to a flood. Four main approaches are used. Simply accepting the financial losses resulting from flooding is one response to flooding, but is not a very positive one. The loss may be borne by individuals or by public funds. Increasingly this approach is used less and less, as the funds set aside are usually not large enough to compensate fully for flood losses. Flood insurance is a second response, and a typical example is the National Flood Insurance Program started in the USA in 1968. The scheme provides low cost insurance for buildings and property and in the first three years 100000 policies were purchased. A similar scheme is available through a New Zealand government backed agency. The third approach is to apply strict land-use planning policies in potential flood areas. The difficulties with this approach are in identifying potential flood areas, and in deciding what to do with development which has already taken place in flood hazard zones. The final approach is to improve forecasting in order to give advance warning of floods so that basic action can be taken to reduce damage. Flood forecasting is improving rapidly in its accuracy, and Operation Foresight, which warned of floods on the Mississippi River in 1969, reduced an estimated potential damage bill of $360 million to an actual damage bill of $100 million. A wide range of radar and satellite technology is used in developing better flood forecasts.

All these schemes of flood control, both structural and behavioural, try to influence the various processes generating floods and to modify these processes. In modifying the processes landscapes are changed and man takes a very active role in determining what landscapes look like.

Landscape policies in the Third World

Many of the examples mentioned in the preceding paragraphs are from Europe or North America, but planned landscape change takes place just as commonly in less developed countries. In urban areas comprehensive land-use plans may exist in the rapidly expanding cities in the less developed world. In the rural areas major planned social reforms may be under way, which radically change agricultural and settlement patterns. The planned changes underway in Tanzania have features typical of many countries of the Third World.

Prior to 1967, attempts to improve rural living standards focused on resettlement programmes with the rehousing and concentration of population in villages. After 1967 the plans were more comprehensive and aimed to change almost all aspects of rural society by the creation of *ujamaa* villages, cooperative communities in which people live and work for cooperative rather than personal goals. The creation of these new settlements involved:

1   Collectivisation and concentration of the population into nucleated *ujamaa* villages

2 Communal ownership of the land
3 Formation of village governments with power to plan and implement local projects
4 Provision of social services such as schools, dispensaries and water supply

By 1969 there were 180 of these villages containing about 60000 people. Although initial progress was relatively slow, the five year plan for Tanzania which began in 1969 positively encouraged rural development and progress quickened. There were 5628 settlements involving over two million people by 1973. The rural development schemes were phased with, first, the physical building of the village, then the introduction of cooperative organisations and thirdly the move towards a fully cooperative society. A series of regional planning communities were given the responsibility of implementing the phased plan and the target was completion of phase 1 by 1976. By late 1974 all the rural population had been collected into villages or small areas of dispersed but very concentrated rural settlement. From Fig. 17.2 it can be appreciated that major changes took place in the landscape as a result of this resettlement programme. Inevitably there were problems in such a rapid relocation of the rural community, but by allowing individual plots of a half hectare for each home, as well as allocating to farmers land in the area surrounding the village, many problems of land redistribution were removed.

The Reregistration and Administration Act of 1975 allowed each of the new villages formally to become a full cooperative. The result of this comprehensive approach to rural planning has been to improve the general standard of living of the rural population at the expense of a fall in living standards for the previously few rich landowners. Basic services, such as essential medical and educational facilities, and important foods are now available to all the rural population. But there is a great shortage of luxuries, with, for example, very few cars even among the relatively well-off sections of society. This is in strong contrast to the pattern seen in many of the developing countries, where luxury goods are only available for the small but significant number of rich people.

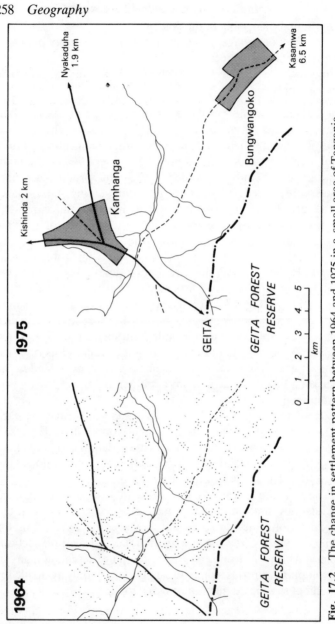

Fig. 17.2 The change in settlement pattern between 1964 and 1975 in a small area of Tanzania

The need for accurate forecasting

Land-use and social planning has become commonplace over the last fifty years, not only for flood control or rural development as shown in this chapter. Landscapes show the imprint of these planning activities far more now than three or four decades ago. It is becoming increasingly difficult to find inhabited regions where governmental planning agencies are not responsible for some of the current processes of landscape change. Although our knowledge of landscape processes has improved in the last fifty years, enabling planning to be more sensitive to existing landscape processes, disasters can and do happen when plans are drawn up for landscape change which effectively cannot occur, given the existing landscape patterns and processes.

In order that planning activity and planned landscape change can occur, it is often necessary to *forecast* how landscape processes will respond to planned changes. Increasingly in recent years geographers have become involved in trying to forecast how the landscape will change *if* such and such happens, or *if* some other policy is adopted. What will happen if a motorway is built along route *x*? What will happen if policy *y* is adopted to stimulate inner city areas? How will building shopping centre *z* affect shops in the city centre? Generally three types of approach have been used for the forecasting exercise. First there is the method of extending trends, *trend forecasting*, from the past to the future. With this method we may extend past cycles into the future or simply extend into the future the past line of development. For short period projections this method is often quite accurate, but in earlier chapters we have seen how both stage and event models may be used to explain landscape processes and in neither case is the future a simple trend extension of the past. A second type of forecast method must therefore be used in which the total process is considered in an integrated and analytical way and the future is worked out by estimating how the individual components will change, called *system forecasting*. Weather forecasts are of this type. An assessment is made of the processes in depressions, anticyclones, fronts and other features, then forecasts are made as to how they will interact and what the resulting weather will be. The amount of analysis in this method of forecasting may vary tremendously from the very analytical computer-dependent weather forecasts to the recent popular forecasts of the effects of micro-electronics on cities, which in many cases are 'crystal ball guessing' with little or no analytical back-up. The trend forecast and the integrated system forecast both depend on some sort of analysis of the process being forecast. A third method, *analogue forecasting*, is to compare the process under study with other processes, establish correlations and associations (although there may not be *direct* links), see how these related processes are changing and then make a forecast of the main process. This approach has tended to be more popular in economics than in geography. It is fraught with difficulty because we are effectively forecasting processes without really understanding how they work.

All these forecasting activities are difficult and the geographer, like the weather forecaster, can be very wrong in the forecasts he makes. But in order to make a useful contribution to the planning of man's environment and landscape, geographers have to be willing to stick their necks out and forecast the future. Having learnt how to describe, analyse and explain the landscape as it exists today, the geographer is perhaps in a better position than anyone to forecast how the landscape of the future will look. Certainly most modern geographers feel that geography has a moral obligation to try.

Further Reading

A general note on atlases, sources and textbooks

Now that the geographer is no longer expected to be a walking gazetteer, memorising all the countries, their capitals, rivers, mountains, exports and imports, it is important that the student of geography has access for reference to a good atlas, large scale maps and up-to-date statistical material. Several excellent atlases exist, such as:

University Atlas, published by George Philip, London.
New Oxford Atlas, published by Oxford University Press, London.
The New Penguin World Atlas, published by Penguin Books.

The Economic Atlases produced by Oxford University Press are valuable reference sources combining both maps and statistics of products, exports, etc.

Although rather expensive for personal use, the *United Nations Statistical Yearbook* is a standard source of economic and social statistics available in many public libraries. The *Geographical Digest*, published each year by George Philip and Son, contains useful summary statistics.

Definitions of terms are provided in W. G. More *Dictionary of Geography* (1981), Penguin, Harmondsworth. Concerned only with human geography is R. J. Johnston (Ed.) *The Dictionary of Human Geography* (1982), Basil Blackwell, Oxford.

A brief guide to specific further reading is provided below associated with each chapter. More general texts are listed first, that both amplify the range of material within this book and also provide alternative perspectives on the study of geography and the interpretation of landscape. Dates of publication generally refer to the most recent edition.

A unitary and stimulating introduction to geography is provided in P. Haggett, *Geography: A Modern Synthesis* (1979), Harper and Row, London.

As an introduction to physical geography, copiously illustrated texts are:

A. N. and A. H. Strahler, *Modern Physical Geography* (1978), Wiley, Chichester.

G. Dury, *An Introduction to Environmental Systems* (1981), Heinemann, London.

and to human geography:

R. L. Morill and J. M. Dormitzer, *The Spatial Order* (1979), Duxbury, North Scituate.

R. Abler, J. S. Adams and P. Gould, *Spatial Organization* (1971), Prentice Hall, Englewood Cliffs.

These four books are American in origin. Amongst British books the following are useful, although slightly beyond the introductory level:

J. Rice, *Geomorphology* (1977), Longmans, London.

D. M. Smith, *Human Geography: A Welfare Approach* (1977), Arnold, London.

More traditional approaches are adopted in:

J. O. M. Broek and J. W. Webb, *A Geography of Mankind* (1973), McGraw Hill, London.

K. W. Butzer, *Geomorphology from the Earth* (1978), Harper and Row, London.

A considerable variety of textbooks are available related to the levels of geography taught in schools, and the review pages of the journals *Geography* and *Teaching Geography* should be consulted to discover books in these fields. For the reader who might be intending British 16+ public examinations the following texts may be useful.

In physical geography:

K. Hilton, *Process and Pattern in Physical Geography* (1979), University Tutorial Press, London.

M. J. Bradshaw, A. J. Abbott and A. P. Gelsthorpe, *The Earth's Changing Surface* (1978), Hodder and Stoughton, London.

D. and V. Weyman, *Landscape Processes: An Introduction to Geomorphology* (1977), Allen and Unwin, London.

In human geography:

M. G. Bradford and W. A. Kent, *Human Geography* (1977), Oxford University Press, London.

J. A. Dawson and D. Thomas, *Man and His World* (1975), Nelson, London.

C. Whynne-Hammond, *Elements of Human Geography* (1979), George Allen and Unwin, London.

Specifically dealing with the techniques of geography are:

K. Selkirk, *Pattern and Place* (1982), Cambridge University Press, London.

R. Hammond and P. S. McCullagh, *Quantitative Techniques in Geography* (1974), Oxford University Press, London.

A. Guest, *Advanced Practical Geography* (1975), Heinemann, London.

Of the monthly magazines, *Geographical Magazine* provides a wealth of examples of landscape development around the world and *Scientific American* has frequent articles of geographical interest. *The Economist* provides a regular source of valuable material on the background to landscape change.

Chapter 1

The Making of the English Landscape is the subject of an introduction and series of books edited by W. G. Hoskins (Hodder and Stoughton). Individual counties are the subject of specific volumes, for example, R. Newton, *The Northumberland Landscape* (1972) Hodder and Stoughton, London. These county volumes deal with the landscape change associated with particular industries such as the iron industry mentioned in this chapter.

A wider view is provided in A. R. H. Baker and J. B. Harley, (eds), *Man Made the Land* (1973), David and Charles, Newton Abbot.

More closely argued are the contributions in H. C. Darby, (ed.), *A New Historical Geography of England* (1973), Cambridge University Press, London.

Chapter 2

An introductory discussion of resource definition is provided in the early chapters of J. H. Paterson, *Land, Work and Resources* (1976), Arnold, London.

Specifically for natural resources see I. G. Simmons, *The Ecology of Natural Resources* (1981), Arnold, London.

Techniques for the description of regions, nodes and networks are provided in D. Smith, *Patterns of Human Geography* (1977), Penguin, Harmondsworth.

Chapter 3

The environmental system approach to physical geography is developed in the textbook by J. J. Hidore, *Physical Geography: Earth Systems* (1974), Scott, Foresman, Brighton.

The application of these theories of systems is illustrated in K. E. Sawer, *Landscape Studies* (1975), Edward Arnold, London.

The subsystems in the atmosphere are considered in a general way in:

T. J. Chandler, *The Air Around Us* (1967), Aldus Books, London,
T. F. Gaskell and M. Morris, *World Climate* (1979), Thames and Hudson, London,

and in a more analytical fashion in H. Riehl, *Introduction to the Atmosphere* (1978), McGraw-Hill Kogakusha, London.

A general approach to hydrosphere subsystems is L. B. Leopold, *Water – A Primer* (1974), W. H. Freeman, San Francisco.

More systematic treatments are provided in:

R. C. Ward, *Principles of Hydrology* (1974), McGraw-Hill, London.
D. I. Smith and P. Stopp, *The River Basin* (1978), Cambridge University Press, London.

For more detail on the two geosphere subsystems discussed here see:

A. L. Bloom, *Geomorphology* (1970), Prentice Hall, Englewood Cliffs.
J. T. Wilson, *Continents Adrift and Continents Aground: Readings from Scientific American* (1977), W. H. Freeman, San Francisco.

For detail on the biosphere see *Scientific American (1970) The Biosphere*, W. H. Freeman, San Francisco.

Chapter 4

A useful introduction to map interpretation is P. Speak and A. H. Carter, *Map Reading and Interpretation* (1974), Longman, London.
Map making methods are described in:

J. S. Keates, *Understanding Maps* (1982), Longman, London.
G. R. P. Lawrence, *Cartographic Methods* (1979), Methuen, London.

Somewhat older is the long-established reference guide by F. J. Monkhouse and H. R. Wilkinson, *Maps and Diagrams* (1975), Methuen, London.

For British readers, it is also useful to obtain copies of Ordnance Survey maps at three or four different scales, to look at the different approaches to landscape representation. Consideration of the mapping techniques in any major atlas serves to show most of the methods mentioned in this chapter. It is worthwhile to obtain examples of some foreign maps, for example French or American ones, to compare the work of differing mapping agencies. Foreign maps can be obtained from Stanfords Map Centre in London.

Chapter 5

An introduction to air photograph interpretation is provided in the later chapters of G. E. Dickinson, *Maps and Air Photographs* (1979), Arnold, London. A non-technical treatment is provided by J. K. S. St. Joseph, *The Uses of Air Photography* (1977), A. C. Black, London.

For discussion of a wider range of remote sensing methods, including satellite imagery, see E. C. Barrett and L. F. Curtis, *Introduction to Environmental Remote Sensing* (1976), Chapman and Hall, London.

Air photographs of various types can be obtained from several suppliers and a large selection is available from Hunting Air Services, Weston-Super-Mare. Satellite photographs are available from Eros Data Centre, Sioux Falls, North Dakota, USA.

Chapter 6

A useful introduction to sampling techniques which, despite its title is applicable to all branches of geography is T. L. Burton and G. E. Cherry, *Social Research Techniques for Planners* (1970), Allen and Unwin, London.

A very short but extremely useful guide to survey methods is

C. Dixon and B. Leach, *Sampling Methods for Geographical Research* (1978), Geo-abstracts, Norwich.

The results of the 1981 Census of Population in the UK are available in a variety of formats from the Office of Population Censuses and Surveys and branches of HMSO. Useful reference texts for anyone undertaking social or economic surveys are Office of Population Censuses and Surveys, *Classification of Occupations* (1980), HMSO, London, and *Standard Industrial Classification* (1980), HMSO, London.

Comparable standard classifications exist for most other countries.

Chapter 7
Most of the general physical geography texts mentioned in the note on textbooks have sections on climate and vegetation. A useful introduction to world climates is: D. R. Riley and L. Spolton, *World Weather and Climate* (1974), Cambridge University Press, London, and for vegetation, A. S. Collinson, *Introduction to World Vegetation* (1978), Allen and Unwin, London.

The relationship between soils and vegetation is explored in S. R. Eyre, *Vegetation and Soils* (1975), Arnold, London.

World patterns of population density and several other characteristics are described in J. I. Clarke, *Population Geography* (1965), Pergamon, Oxford.

Chapter 8
General development issues are considered in D. M. Smith, *Where the Grass is Greener* (1979), Penguin, Harmondsworth.

The population characteristics of First and Third World countries are considered in:

G. T. Trewartha, *The More Developed Realm: A Geography of Its Population* (1978), Pergamon, Oxford.

G. T. Trewartha, *The Less Developed Realm: A Geography of Its Population* (1972), Wiley, London.

Associations in agriculture are discussed in:

P. A. R. Newbury, *A Geography of Agriculture* (1980), Macdonald and Evans, Plymouth.

W. W. Morgan and R. J. C. Munton, *Agricultural Geography* (1971), Methuen, London.

A wide-ranging view of arid landscapes is presented in J. Cloudsley-Thompson, *The Desert* (1977), Orbis, London.

Chapter 9

Descriptions of oceanic current flows are contained in major texts on oceanography such as C. A. M. King, *Introduction to Physical and Biological Oceanography* (1975), Arnold, London, particularly Chapter 3.

Population flows within a continent are a major theme in J. I. Clarke and L. A. Kosinski, *Redistribution of Population in Africa* (1981), Heinemann, London.

Migration patterns are explained in L. A. Kosinski and R. M. Prothero, *People on the Move* (1974), Methuen, London.

The pattern of world trade flows is described in C. A. R. Hills, *World Trade* (1981), Batsford, London.

Chapter 10

There are many books on various aspects of cities and urban geography and most use the idea of models against which to measure real landscapes. Valuable approaches are provided by:

H. Carter, *The Study of Urban Geography* (1981), Arnold, London.
R. J. Johnston, *City and Society. An Outline for Urban Geography* (1980), Penguin, Harmondsworth.

The history of urban development is introduced in G. Burke, *Towns in the Making* (1975), Arnold, London.

A comparison of various land capability classifications is presented in D. A. Davidson, *Soils and Land-Use Planning* (1980), Longman, London.

The study of slopes is an integral part of the general geomorphology books mentioned in the note on textbooks.

Chapter 11

The models of industrial location and settlement patterns are analysed in the general human geography books listed in the note on textbooks.

A discussion of volcanic activity is provided in P. Francis, *Volcanoes* (1976), Penguin, Harmondsworth.

An introduction to the model-building approach to geography is R. Minshull, *An Introduction to Models in Geography* (1975), Longman, London.

Chapter 12

Environmental nutrient cycles are considered in most ecology textbooks. Useful accounts from different viewpoints are included in:

G. Dury, *Environmental Systems* (1981), Heinemann, London.

D. and M. Pimentel, *Food, Energy and Society* (1979), Arnold, London.

The environmental and cultural bases of agriculture in Asia are discussed in several regional textbooks such as: B. L. C. Johnson, *Pakistan* (1980), Heinemann, London, and by the same author and publisher *Bangladesh* (1975) and *India* (1979).

Issues associated with poverty and development cycles are central to:

A. B. Mountjoy, *The Third World: Problems and Perspectives* (1978), Macmillan, London.

A. B. Mountjoy, *Industrialization and Developing Countries* (1982), Hutchinson, London.

Chapter 13

A clear analysis of the processes of economic development is provided in A. L. Mabogunje, *The Development Process* (1980), Hutchinson, London.

Case studies are presented in B. W. Hodder, *Economic Development in the Tropics* (1980), Methuen, London.

A clear account of Von Thünen type models is contained in M. Chisholm, *Rural Settlement and Land Use* (1979), Hutchinson, London.

Glacial landscapes are analysed in most geomorphology textbooks. The debate on the causes and number of ice ages is reviewed in:

B. S. John, *The Ice Age* (1977), Collins, London.

N. Calder, *The Weather Machine* (1974), BBC Publications, London.

Chapter 14

A comprehensive text on the diffusion process is L. A. Brown, *Innovation Diffusion* (1981), Methuen, London.

Chapter 15

A general view of the landscape effect of event phenomena is provided in J. Whittow, *Disasters: The Anatomy of Environmental Hazards* (1980), Penguin, Harmondsworth.

A. H. Perry, *Environmental Hazards in the British Isles* (1981), George Allen and Unwin, London.

The rôle of government in landscape change is implicit in G. Manners et al, *Regional Development in Britain* (1980), Wiley, Chichester,
and is more explicit in sections of:

R. Muir and R. Paddison, *Politics, Geography and Behaviour* (1981), Methuen, London.
K. R. Cox, *Location and Public Problems* (1979), Blackwell, Oxford.

Chapter 16
There is a large number of books on pollution of various types. An introduction is provided by:

R. G. Adamson, *Pollution: An Ecological Approach* (1980), Heinemann, London.
R. Mabey, *The Pollution Handbook* (1974), Penguin, Harmondsworth.

A global view is taken in M. W. Holdgate and G. F. White, *Environmental Issues* (1977), Wiley, London.
The responsibilities of society in creating pollution are analysed in C. H. Waddington, *The Man-made Future* (1978), Croom Helm, London.

Chapter 17
Land-use planning activities in Britain, with some European and American comparisons, are thoroughly covered in P. Hall, *Urban and Regional Planning* (1975), Penguin, Harmondsworth.
Useful views of specific policy approaches in land-use planning are:

P. J. Cloke, *Key Settlements in Rural Areas* (1979), Methuen, London.
I. Alexander, *Office Location and Public Policy* (1979), Longman, London.

Responses to flood hazard are reviewed in M. D. Newson, *Flooding and the Flood Hazard in the United Kingdom* (1975), Oxford University Press, London.
Some examples of planning activity in the Third World are provided in S. H. Ominde (ed.), *Studies in East African Geography and Development* (1971), Heinemann, London.

Index